Year 3 Textb

Series Editor: Tony Staneff

Contents

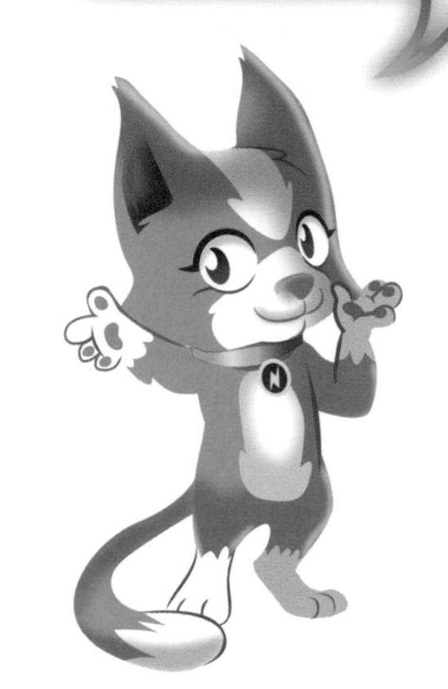

Your teacher will tell you which page you need.

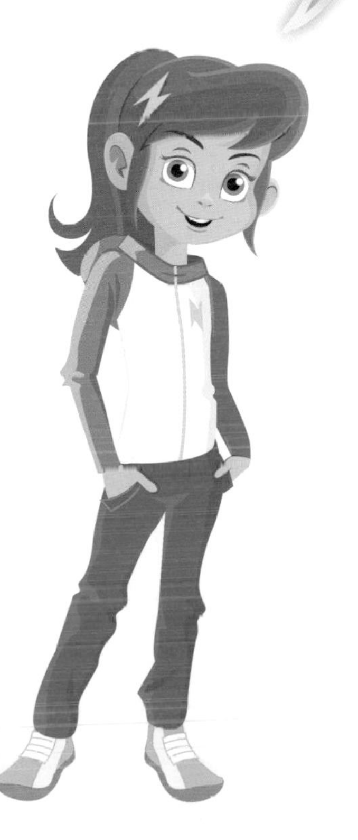
Let's get started!

How to use this book

These pages make sure we're ready for the unit ahead. Find out what we'll be learning and brush up on your skills!

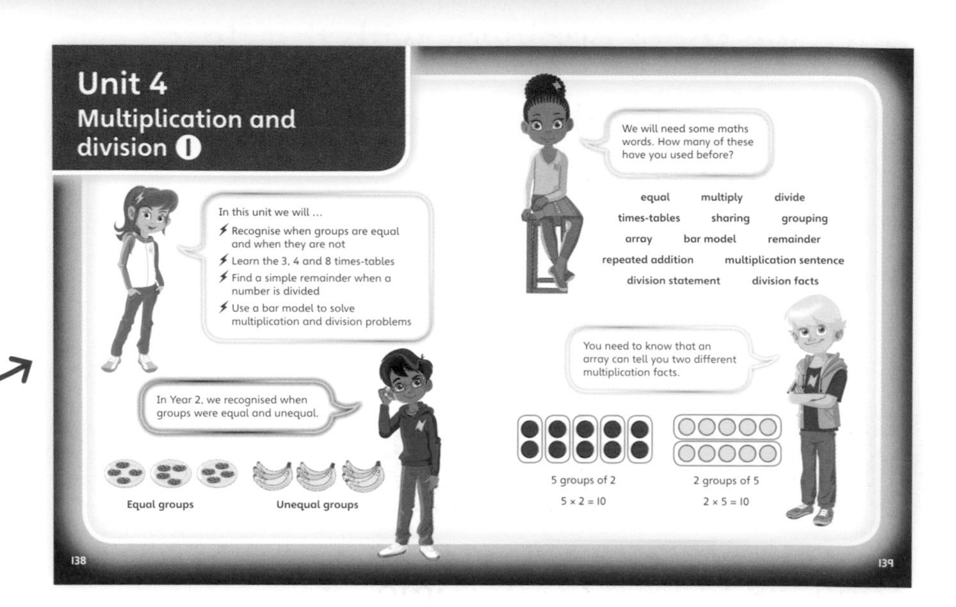

Unit 4
Multiplication and division 1

In this unit we will ...
- Recognise when groups are equal and when they are not
- Learn the 3, 4 and 8 times-tables
- Find a simple remainder when a number is divided
- Use a bar model to solve multiplication and division problems

In Year 2, we recognised when groups were equal and unequal.

Equal groups
Unequal groups

We will need some maths words. How many of these have you used before?

equal, multiply, divide, times-tables, sharing, grouping, array, bar model, remainder, repeated addition, multiplication sentence, division statement, division facts

You need to know that an array can tell you two different multiplication facts.

5 groups of 2
$5 \times 2 = 10$

2 groups of 5
$2 \times 5 = 10$

138 139

Discover

Lessons start with **Discover**.

Here, we explore new maths problems.

Can you work out how to find the answer?

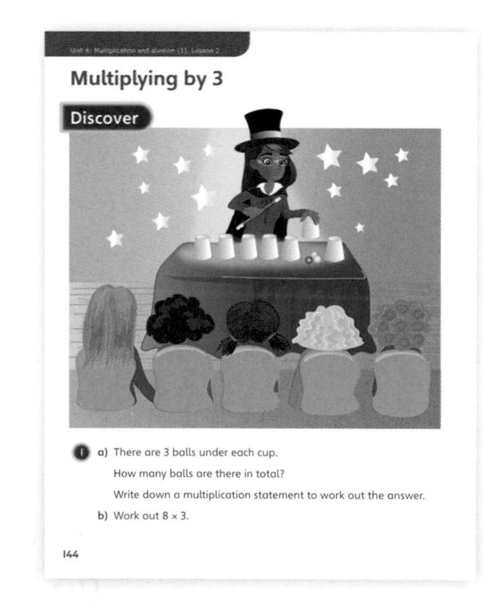

Multiplying by 3

Discover

1 a) There are 3 balls under each cup.
How many balls are there in total?
Write down a multiplication statement to work out the answer.
b) Work out 8×3.

144

Don't be afraid to make mistakes.
Learn from them and try again!

Share

Next, we share our ideas with the class.

Did we all solve the problems the same way? What ideas can you try?

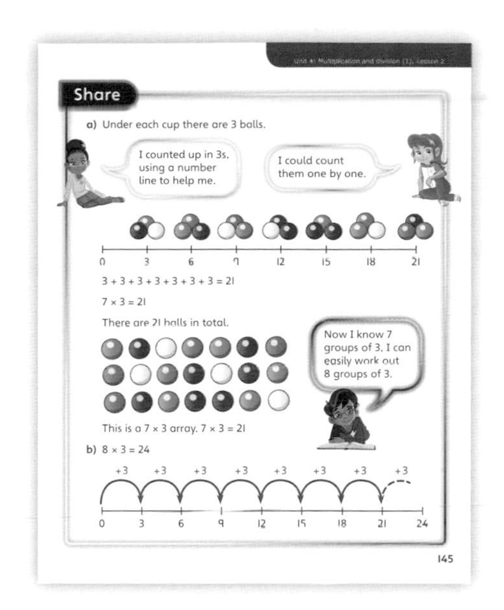

Share

a) Under each cup there are 3 balls.

I counted up in 3s, using a number line to help me.

I could count them one by one.

0 3 6 9 12 15 18 21

3 + 3 + 3 + 3 + 3 + 3 + 3 = 21

7 × 3 = 21

There are 21 balls in total.

Now I know 7 groups of 3. I can easily work out 8 groups of 3.

This is a 7 × 3 array. 7 × 3 = 21

b) 8 × 3 = 24

+3 +3 +3 +3 +3 +3 +3 +3

0 3 6 9 12 15 18 21 24

145

Think together

Then we have a go at some more problems together. Use what you have just learnt to help you.

We'll try a challenge too!

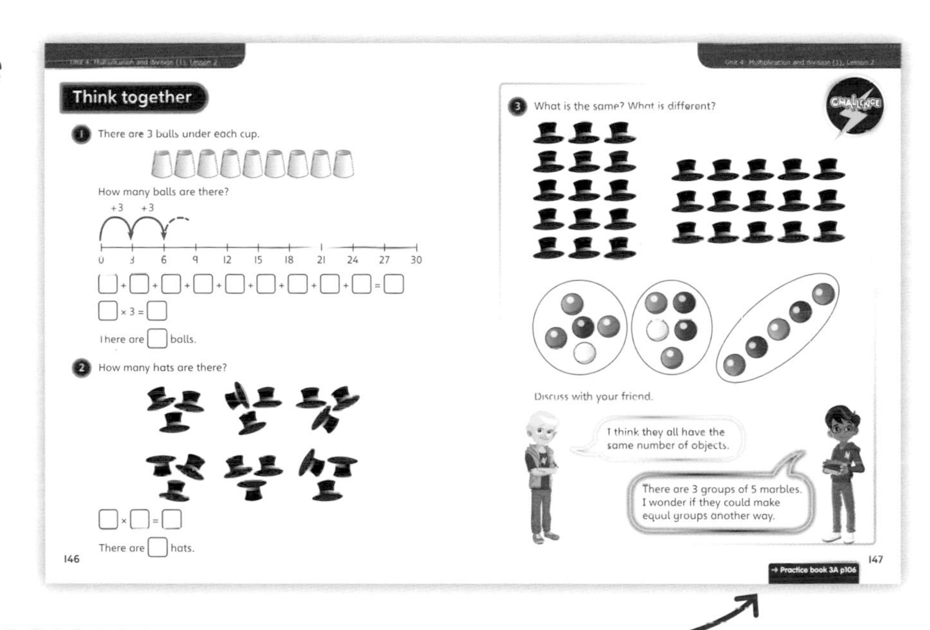

Think together

1. There are 3 balls under each cup.

How many balls are there?

+3 +3

0 3 6 9 12 15 18 21 24 27 30

☐ + ☐ + ☐ + ☐ + ☐ + ☐ + ☐ + ☐ + ☐ = ☐

☐ × 3 = ☐

There are ☐ balls.

2. How many hats are there?

☐ × ☐ = ☐

There are ☐ hats.

146

3. What is the same? What is different?

Discuss with your friend.

I think they all have the same number of objects.

There are 3 groups of 5 marbles. I wonder if they could make equal groups another way.

→ Practice book 3A p106

147

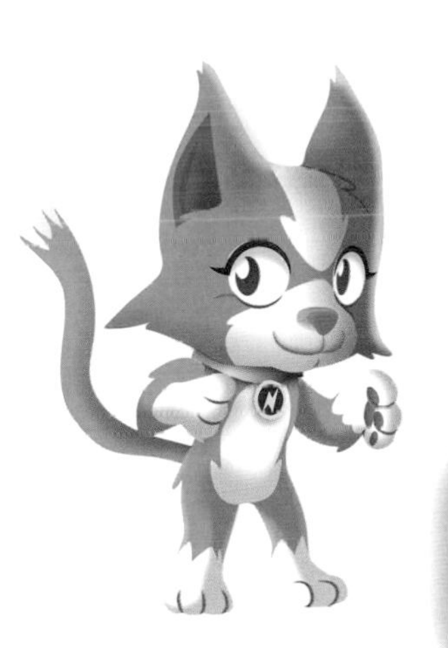

This tells you which page to go to in your **Practice Book.**

At the end of each unit there's an **End of unit check.** This is our chance to show how much we have learnt.

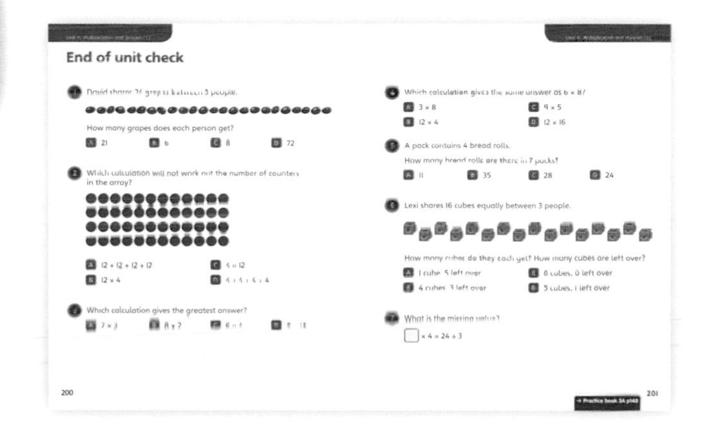

End of unit check

How many grapes does each person get?

A 21 B 6 C 8 D 72

2. Which calculation will not work out the number of counters in the array?

A 12 + 12 + 12 + 12

B 12 × 4

3. Which calculation gives the greatest answer?

A 7 × 3

4. Which calculation gives the same answer as 6 × 8?

A 3 × 8 C 9 × 5

B 12 × 4 D 12 × 16

5. A pack contains 4 bread rolls.

A 11 B 35 C 28 D 24

6. Lexi shares 16 cubes equally between 3 people.

How many cubes do they each get? How many cubes are left over?

☐ × 4 = 24 ÷ 3

200

→ Practice book 3A p148

201

Unit 1
Place value within 1,000

In this unit we will ...

- Count in 100s
- Partition a number in 100s, 10s and 1s
- Find 100, 10 and 1 more or less
- Compare and order numbers up to 1,000
- Count in 50s

In Year 2 we used place value grids to organise our work. What number does this show?

T	O

We will need some maths words.
How many of these can you remember?

hundreds (100s) **tens (10s)**

ones (1s) **place value** **more**

less **greater than (>)** **less than (<)**

equal to **order** **compare**

estimate **exchange**

We will also use part-whole models and number lines.

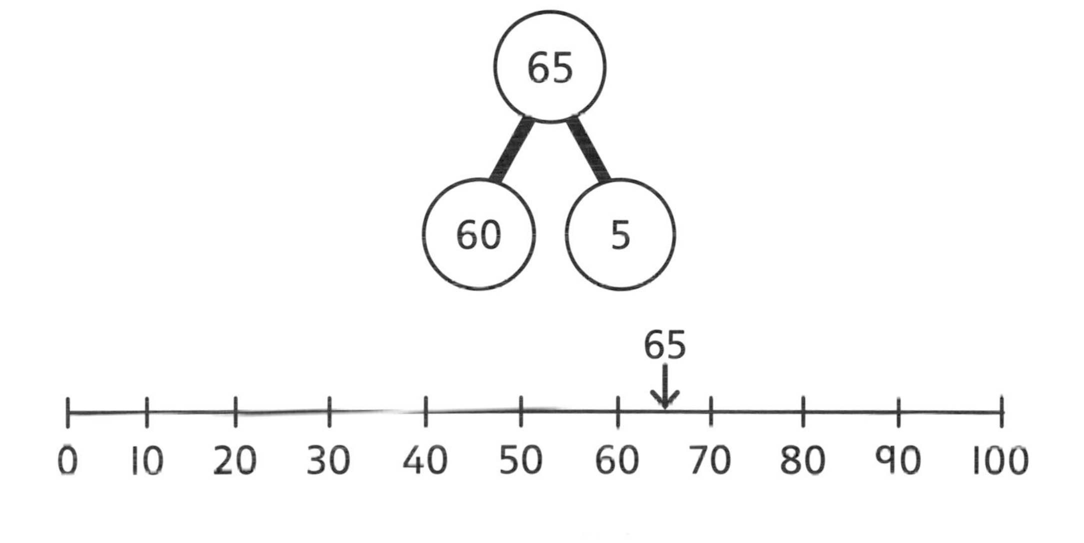

Counting in 100s

Discover

1 **a)** Count the dice on the ground.

Is this the correct amount?

b) How many dice are there in total?

Share

a) We can count the dice.

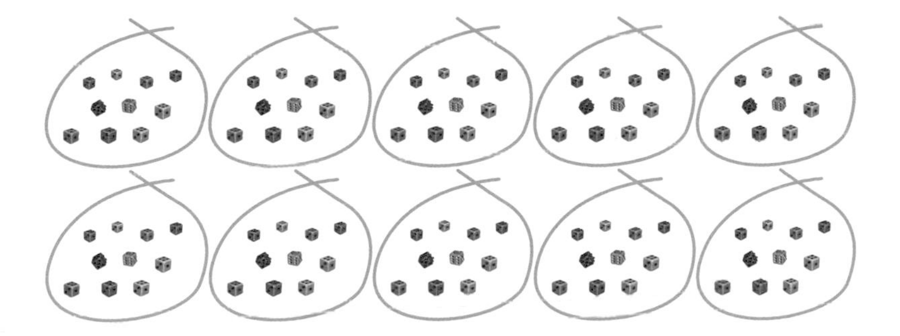

I made groups of 10.

I put them in rows of 10 and then counted in 10s. This looks like a 100 square.

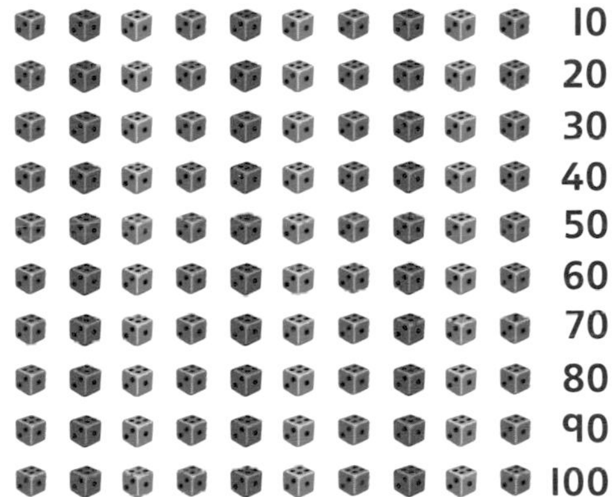

There are 100 dice on the ground. This is the correct amount.

b) There are 100 dice in each jar.

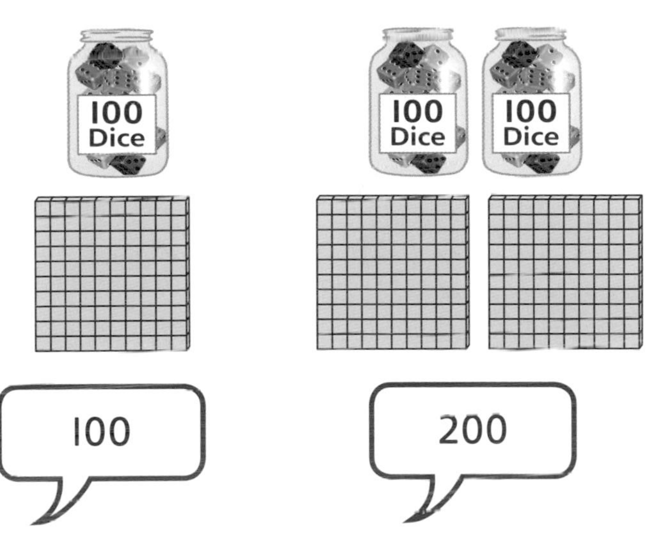

100 Dice 100 Dice 100 Dice

300

There are 300 dice in total.

Think together

Each jar contains 100 counters.

How many are there in each row?

		0	zero
100 Counters		100	one hundred
100 Counters 100 Counters		200	two hundred
100 Counters 100 Counters 100 Counters		300	three hundred
100 Counters 100 Counters 100 Counters 100 Counters			
100 Counters 100 Counters 100 Counters 100 Counters 100 Counters			
100 Counters 100 Counters 100 Counters 100 Counters 100 Counters 100 Counters			
100 Counters 100 Counters 100 Counters 100 Counters 100 Counters 100 Counters 100 Counters			
100 Counters 100 Counters 100 Counters 100 Counters 100 Counters 100 Counters 100 Counters 100 Counters			

2 What are the missing numbers?

a)

0	100	200	300	☐	☐	600	700

b)

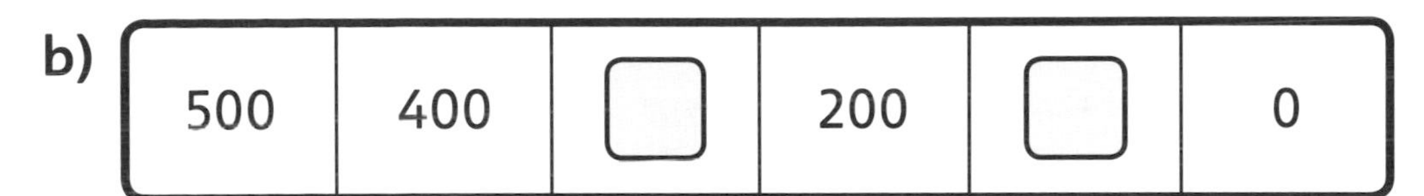

500	400	☐	200	☐	0

c) ☐, 600, ☐, 800, ☐

3 How many marbles are there?

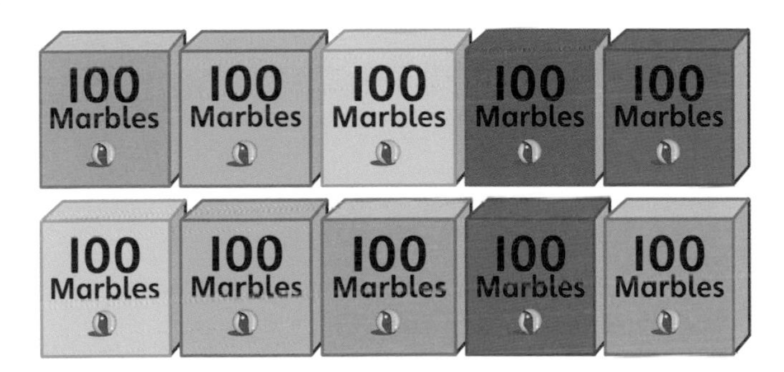

Write your number in numerals and words.

There are ☐ marbles.

There are ________________ marbles.

I know what comes after nine hundred. I think it must be ten hundred.

I think there is another name for this. I wonder what it is.

→ Practice book 3A p6

Representing numbers to 1,000

Discover

There are 235 children in our school.
235
200
30
5

1 **a)** Use base 10 equipment to represent 235.

Show this on a part-whole model.

b) What numbers are represented here?

H	T	O

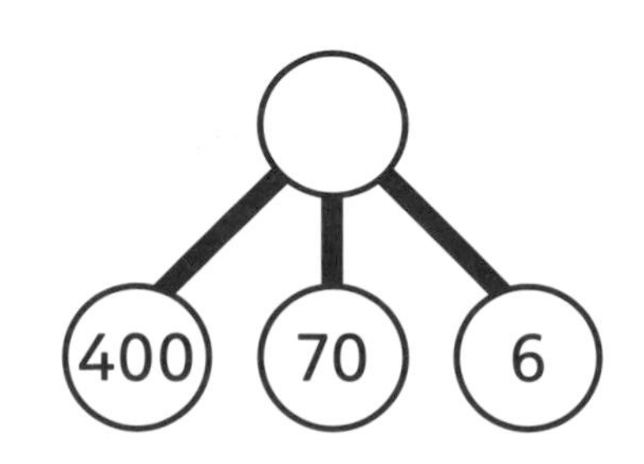

Share

a) There are 235 children in the school.

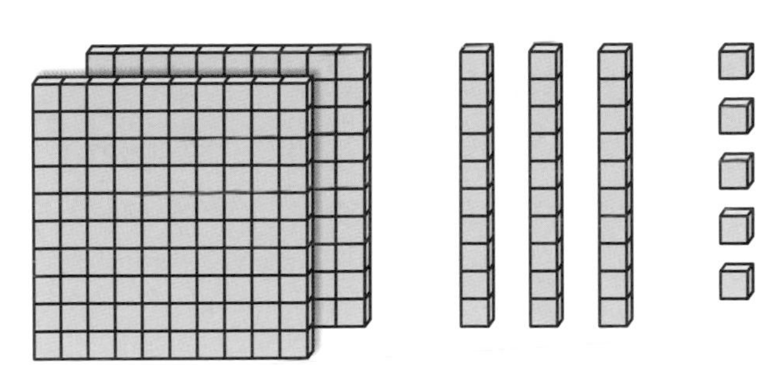

I can represent each child with base 10 equipment.

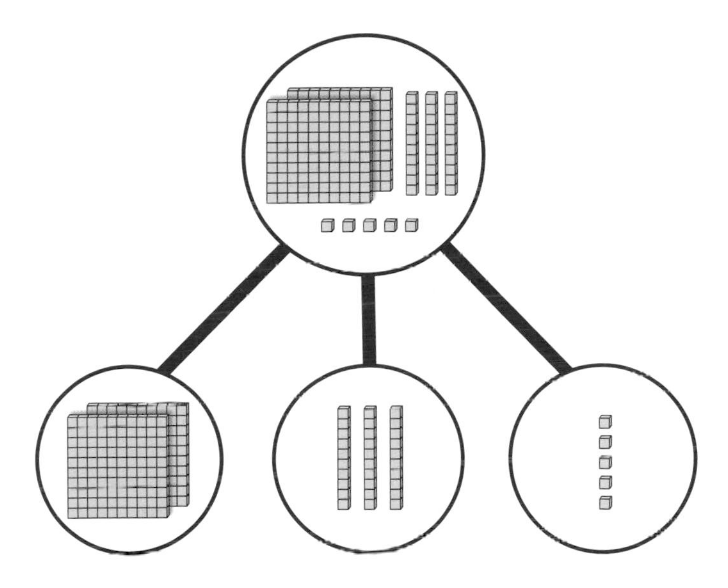

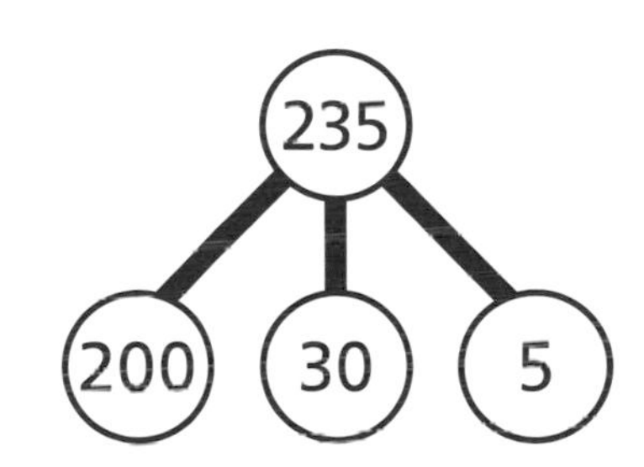

b)

H	T	O

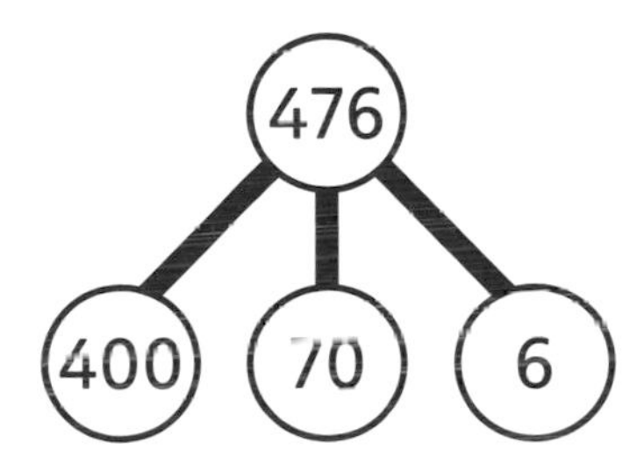

This is 124 (one hundred and twenty-four).

This is 476 (four hundred and seventy-six).

Think together

1 **a)** Represent these numbers using base 10 equipment.

There are 365 days this year.

I have 130 swap cards at home.

b) Show each number on a part-whole model.

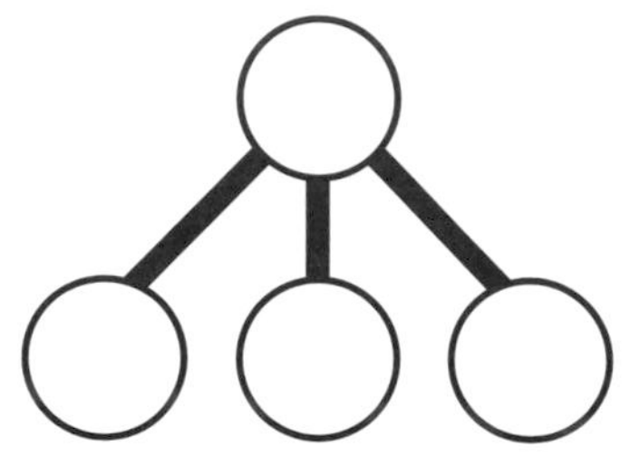

2 What numbers are represented?

a)

H	T	O

b)

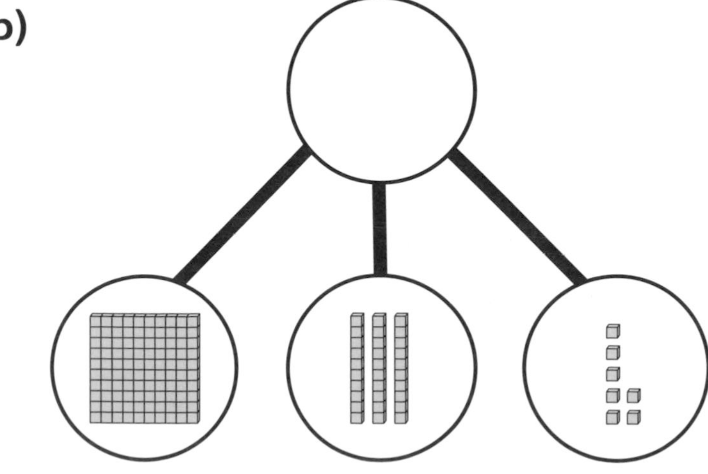

c)

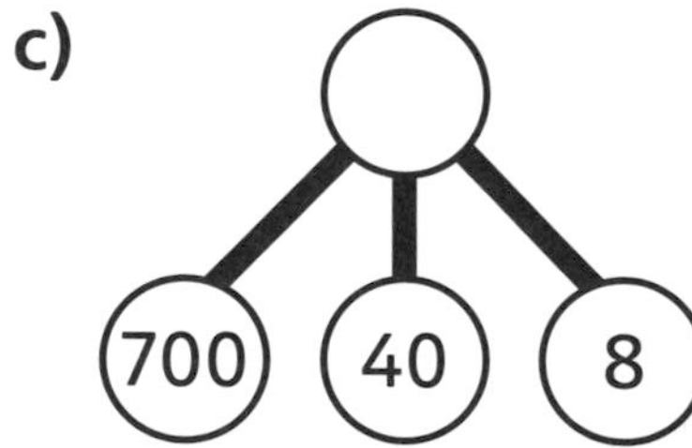

CHALLENGE

3 What mistakes have Andy and Aki made?

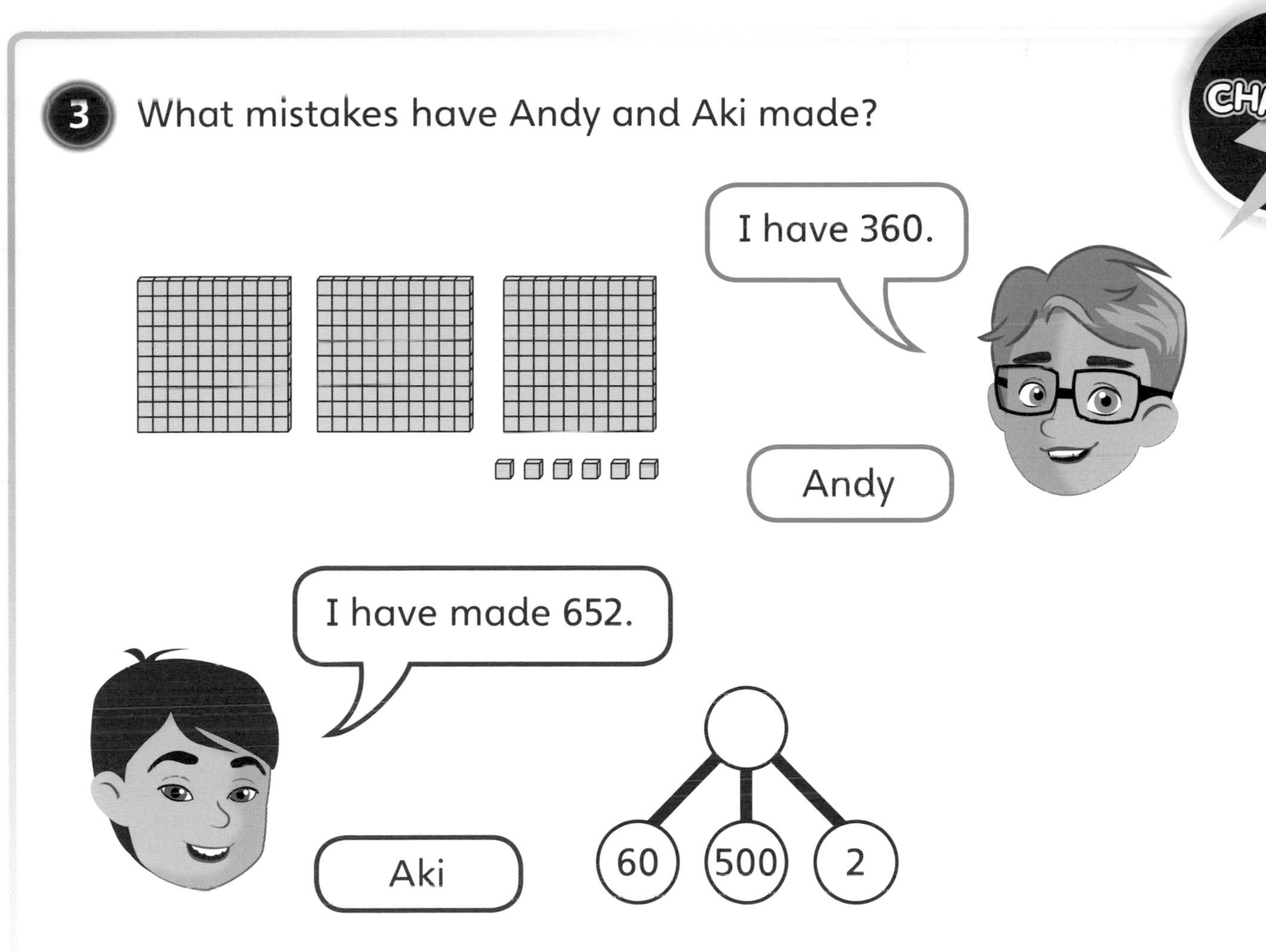
I have 360.
Andy
I have made 652.
Aki
60
500
2

What numbers have Andy and Aki made?

I think they have mixed their numbers up. But which ones?

Use some base 10 equipment to make your own numbers. Challenge your friend.

→ Practice book 3A p9

100s, 10s and 1s 1

Discover

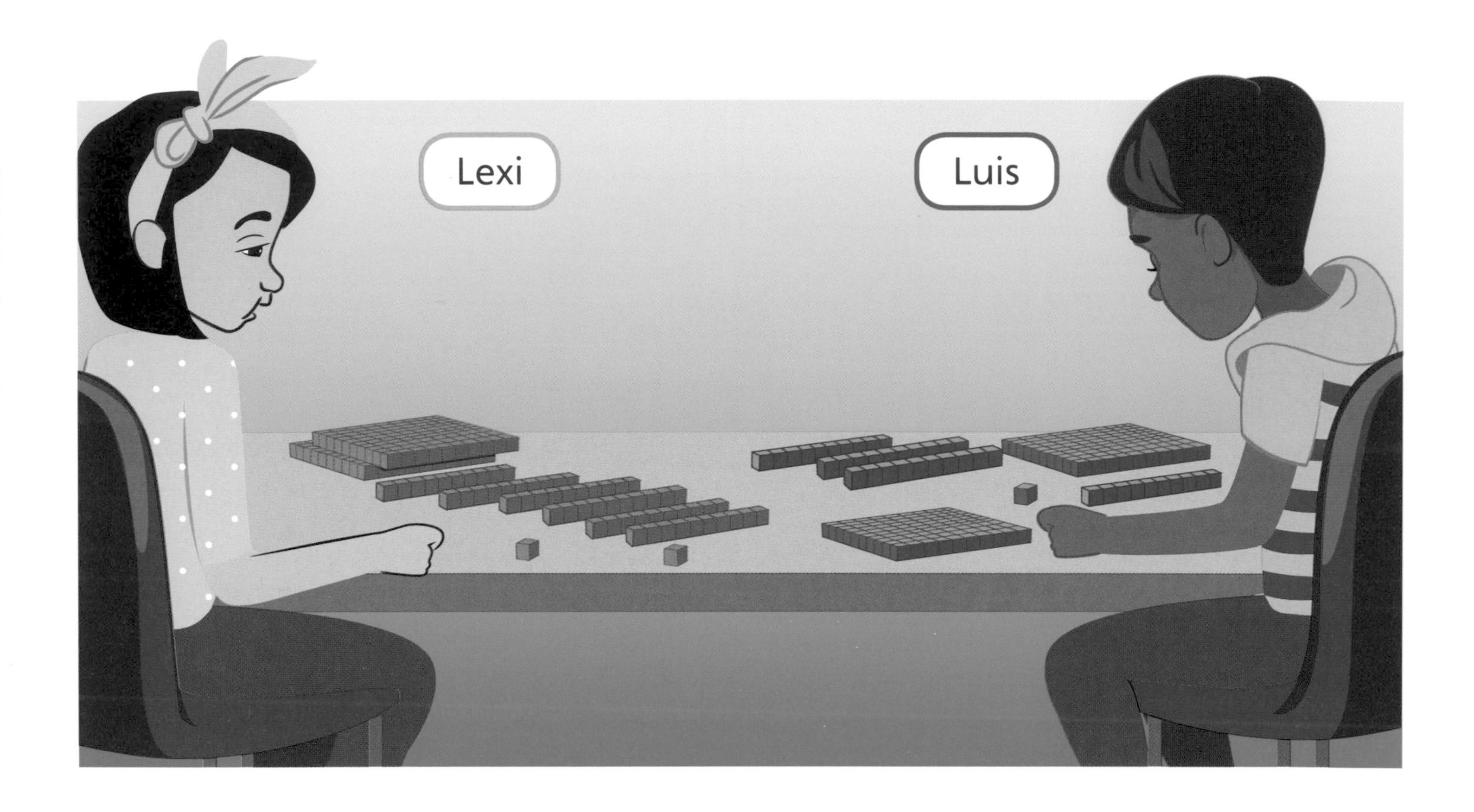

1 a) What numbers have Luis and Lexi made?

Write the numbers in words and numerals.

b) Make the numbers 358 and 430 using base 10 equipment.

Organise your work using a place value grid.

Share

a)

For Luis's number I will count the 100s first, then the 10s and then the 1s.

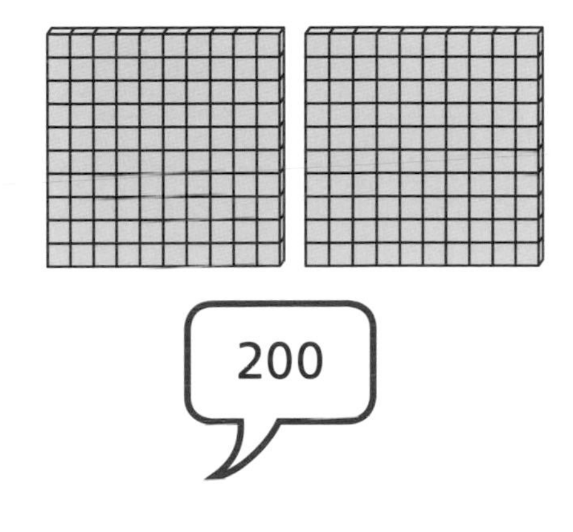

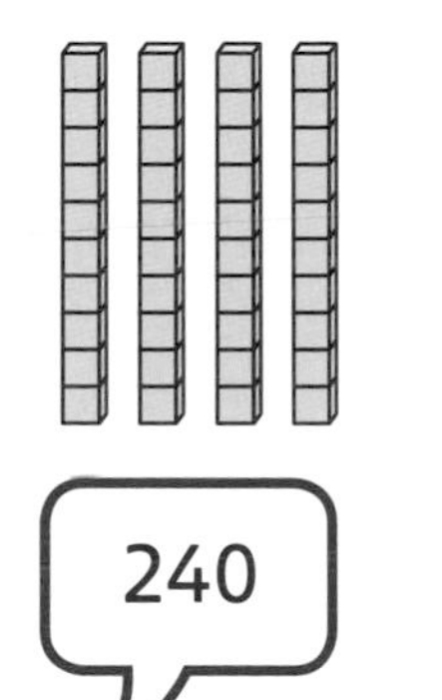

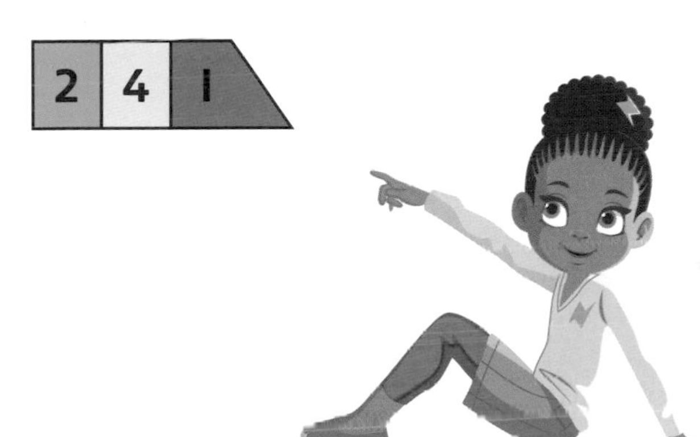

I can show the number using place value cards.

Luis has made the number 241 (two hundred and forty-one).

Lexi has made the number 262 (two hundred and sixty-two).

262 is 2 hundreds, 6 tens and 2 ones.

H	T	O
2	6	2

b)

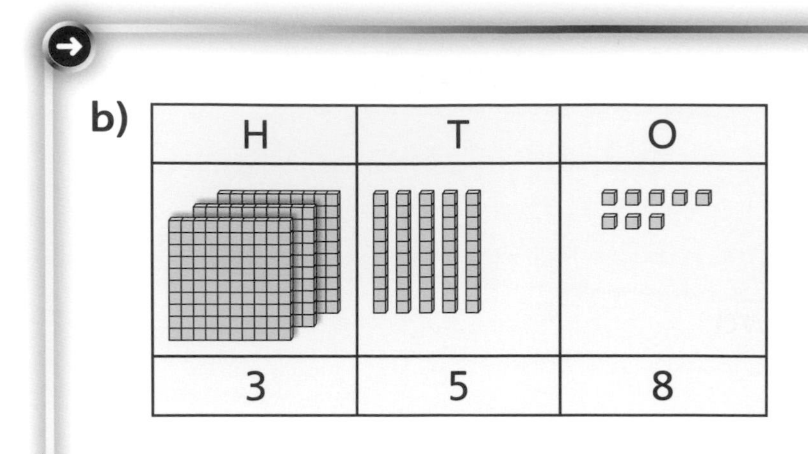

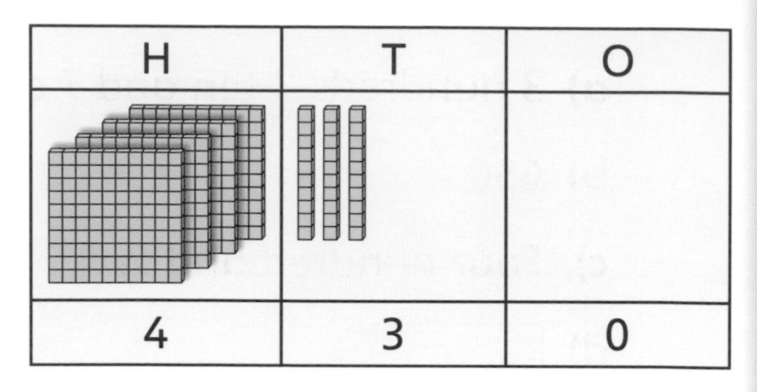

Think together

1 Luis and Lexi each made another number.

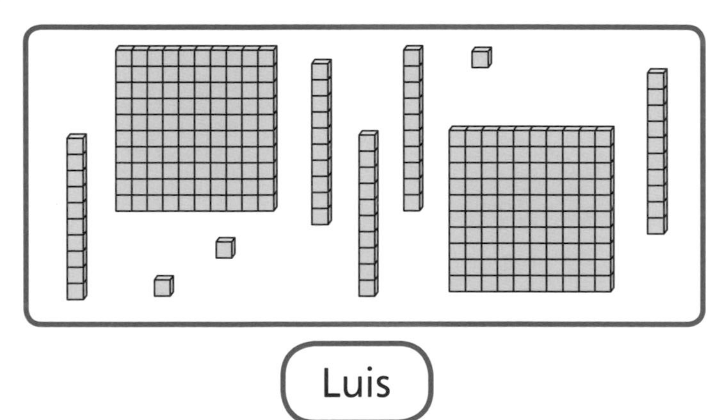

Luis

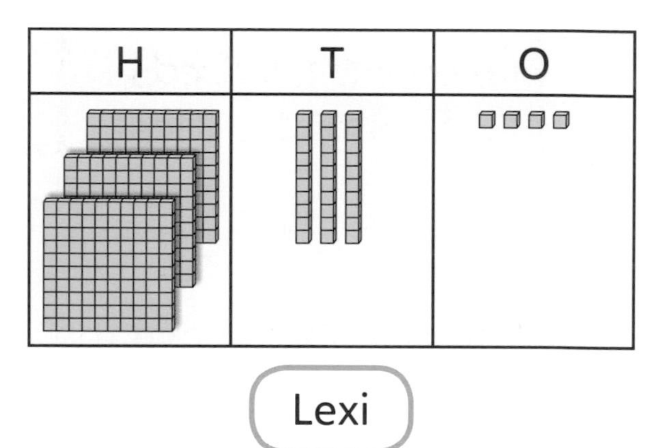

Lexi

a) How many 100s, 10s and 1s did Luis use?

What number did Luis make?

Luis used ☐ hundreds, ☐ tens and ☐ ones.

Luis made the number ☐.

b) What number did Lexi make?

Lexi used ☐ hundreds, ☐ tens and ☐ ones.

Lexi made the number ☐.

2 Represent each of these numbers using base 10 equipment.

a) 3 hundreds, 1 ten and 7 ones

b) 650

c) Four hundred and seventy-two

d) | 7 | 0 | 2 |

3 Luis writes his number like this:

532 is 5 hundreds, 3 tens and 2 ones

Lexi writes her number like this:

262 = 200 + 60 + 2

What numbers have the rest of the class made?

I think two of them have made the same number.

Emma

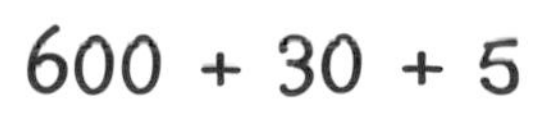

Bella

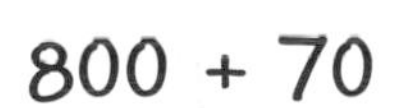

Danny

4 hundreds, 5 tens and 7 ones

Mo

→ Practice book 3A p12

100s, 10s and 1s 2

Discover

1 a) Toshi needs to plant 215 flower bulbs.

How many boxes of 100 bulbs will he need?

How many boxes of 10 bulbs?

How many single bulbs?

b) Use place value counters to represent 215 on a place value grid.

Share

a) Toshi needs to plant 215 flower bulbs.

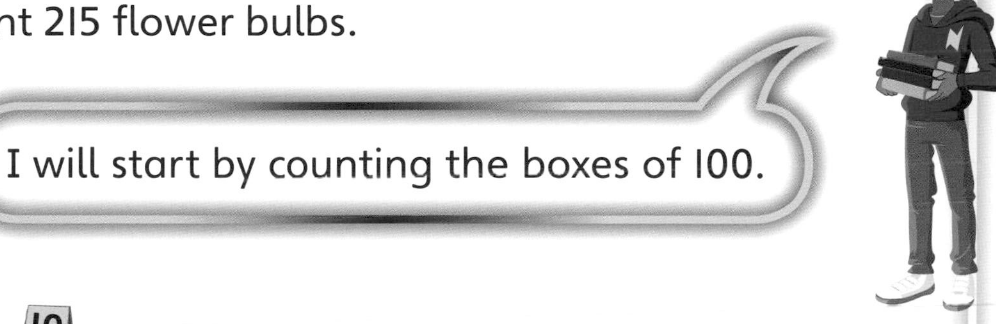

I will start by counting the boxes of 100.

I will use a part-whole model to help me.

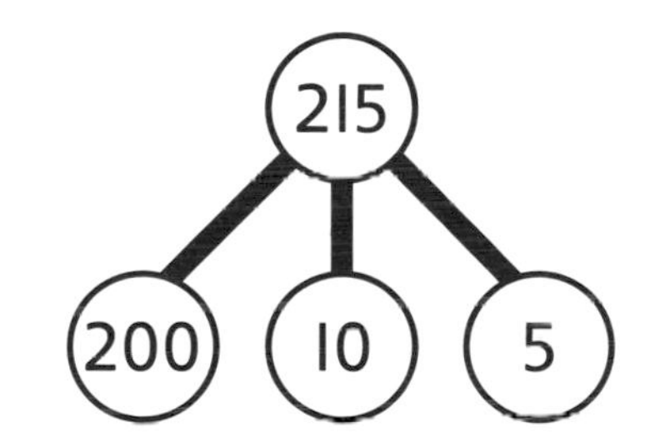

Toshi needs 2 boxes of 100 bulbs, 1 box of 10 bulbs and 5 single bulbs.

b)

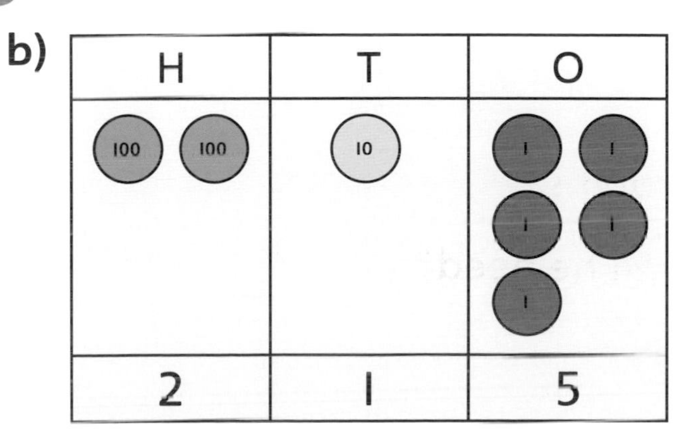

H	T	O
100 100	10	1 1 1 1 1
2	1	5

A place value grid allows us to organise our work into columns. We can quickly see what the numbers are.

Think together

1 Faisal needs to plant 392 sunflower seeds.

How many of each pack does Faisal need?

Faisal needs ☐ packs of 100,

☐ packs of 10 and

☐ single seeds.

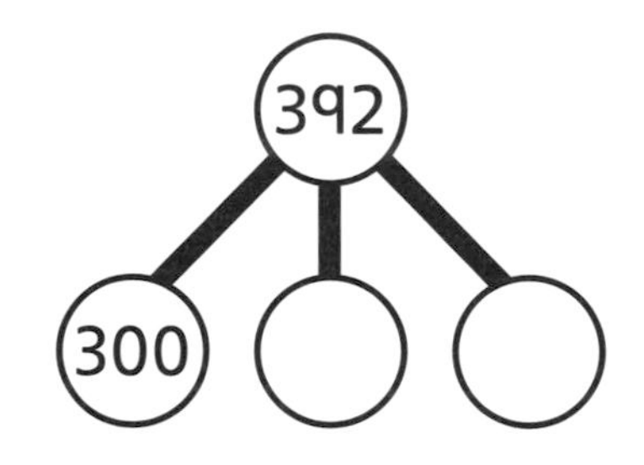

2 What numbers are represented on these place value grids?

a)

H	T	O
100 100 100 100	10	1 1 1 1 1

b)

H	T	O
100 100 100 100 100 100	10 10 10	

c)

H	T	O
100 100 100 100 100		1 1 1

d)

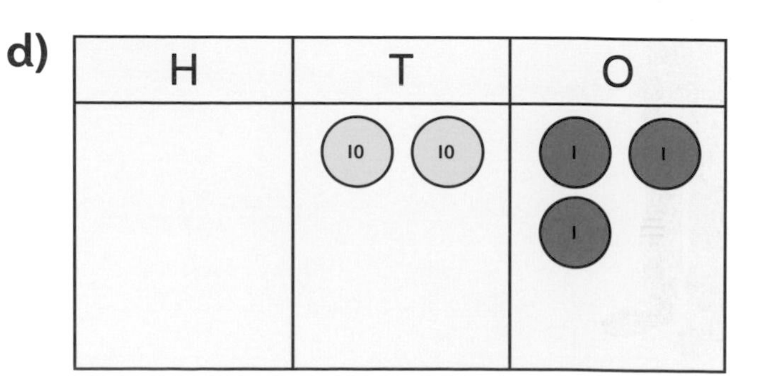

H	T	O
	10 10	1 1 1

a) Use counters to represent each number on a place value grid.

600 + 20 + 7

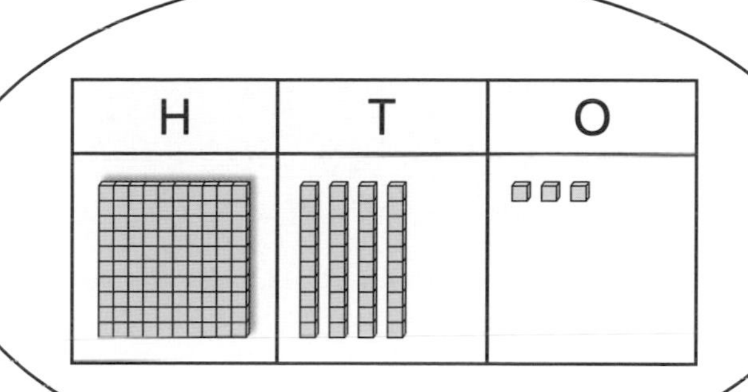

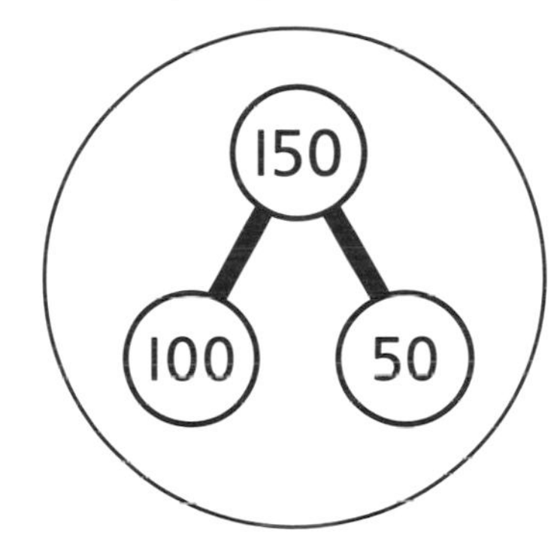

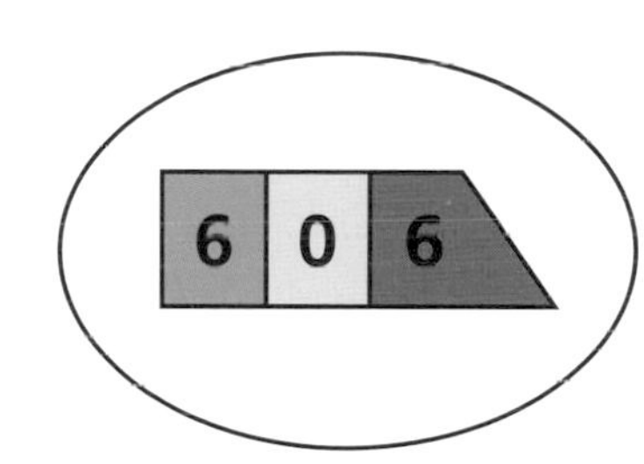

b) Meg made one of the above numbers.
Which number did she make?

H	T	O

I don't think this is a number. There are no numbers on the counters.

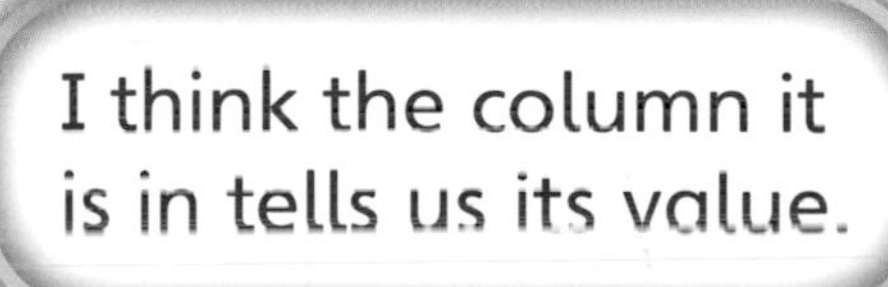

I think the column it is in tells us its value.

→ Practice book 3A p15

The number line to 1,000 ①

Discover

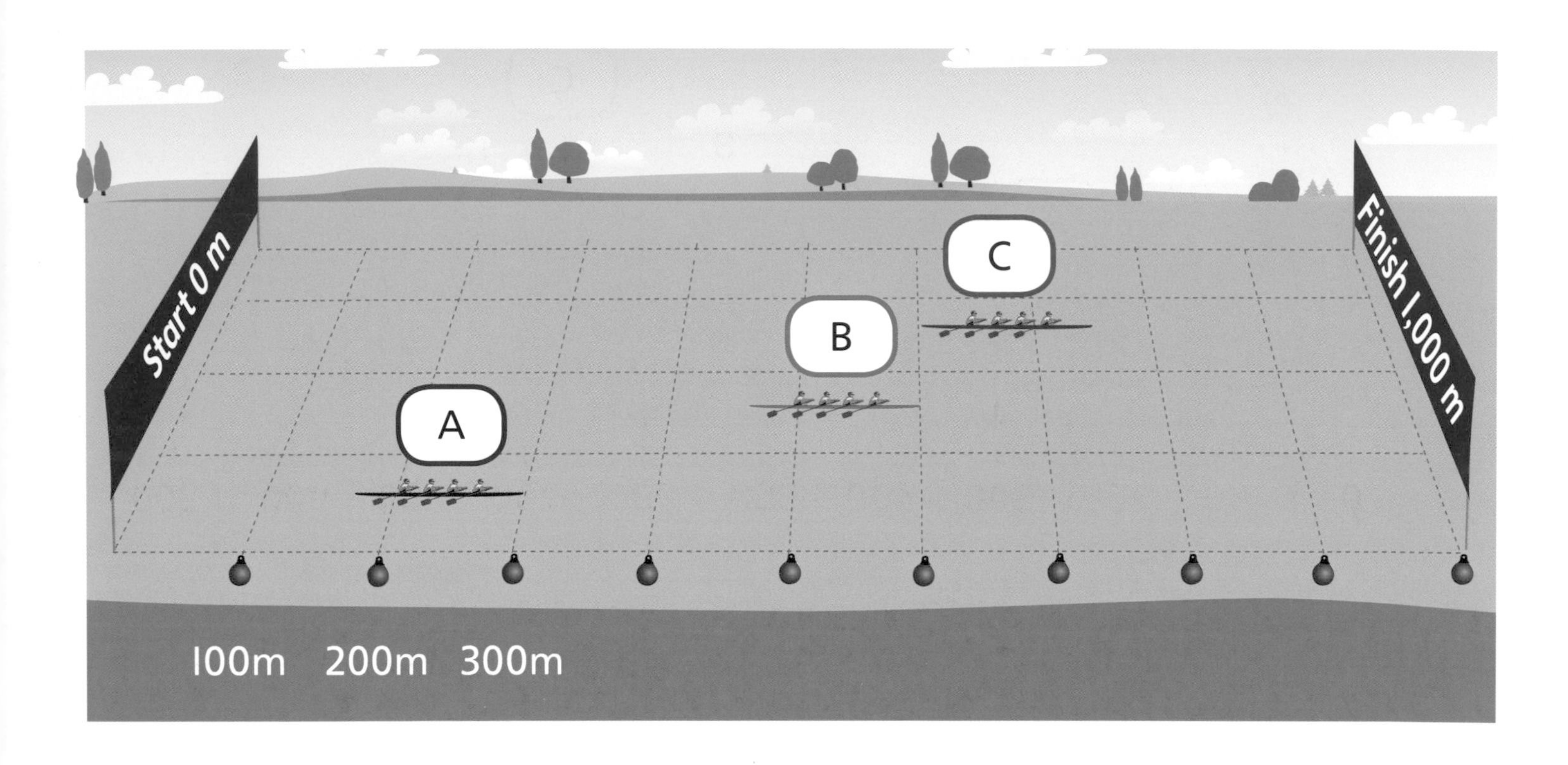

1 **a)** This is a boat race.

How far has boat A travelled?

How far has boat B travelled?

Estimate how far boat C has travelled.

b) Another boat, boat D, has travelled 900 metres.

Where will this boat be?

Share

a) I used a number line to help me. The number line goes up in 100s, from 0 to 1,000.

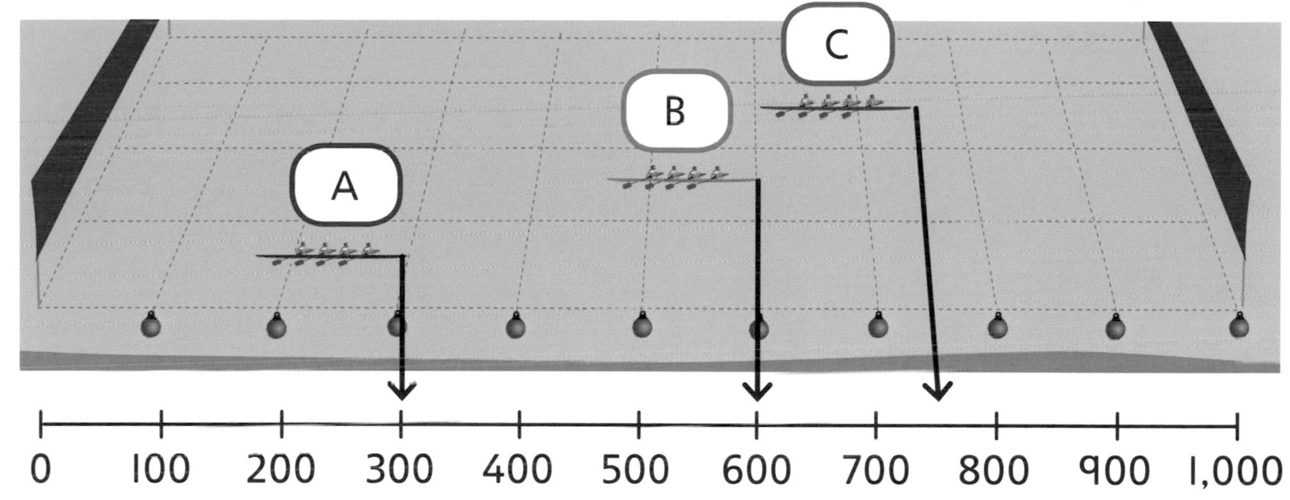

Boat A has travelled 300 metres.

Boat B has travelled 600 metres.

Boat C has travelled about 750 metres.

750 lies half-way between 700 and 800.

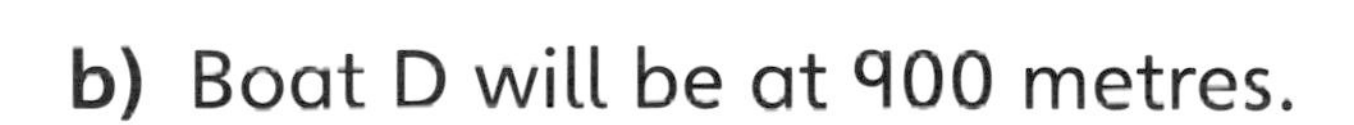

b) Boat D will be at 900 metres.

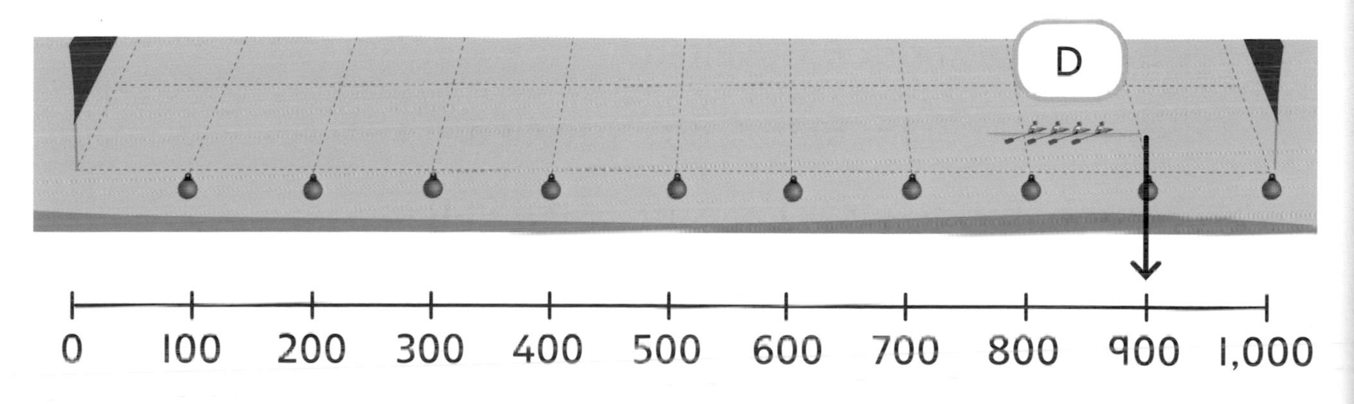

Think together

1 This shows another boat race.

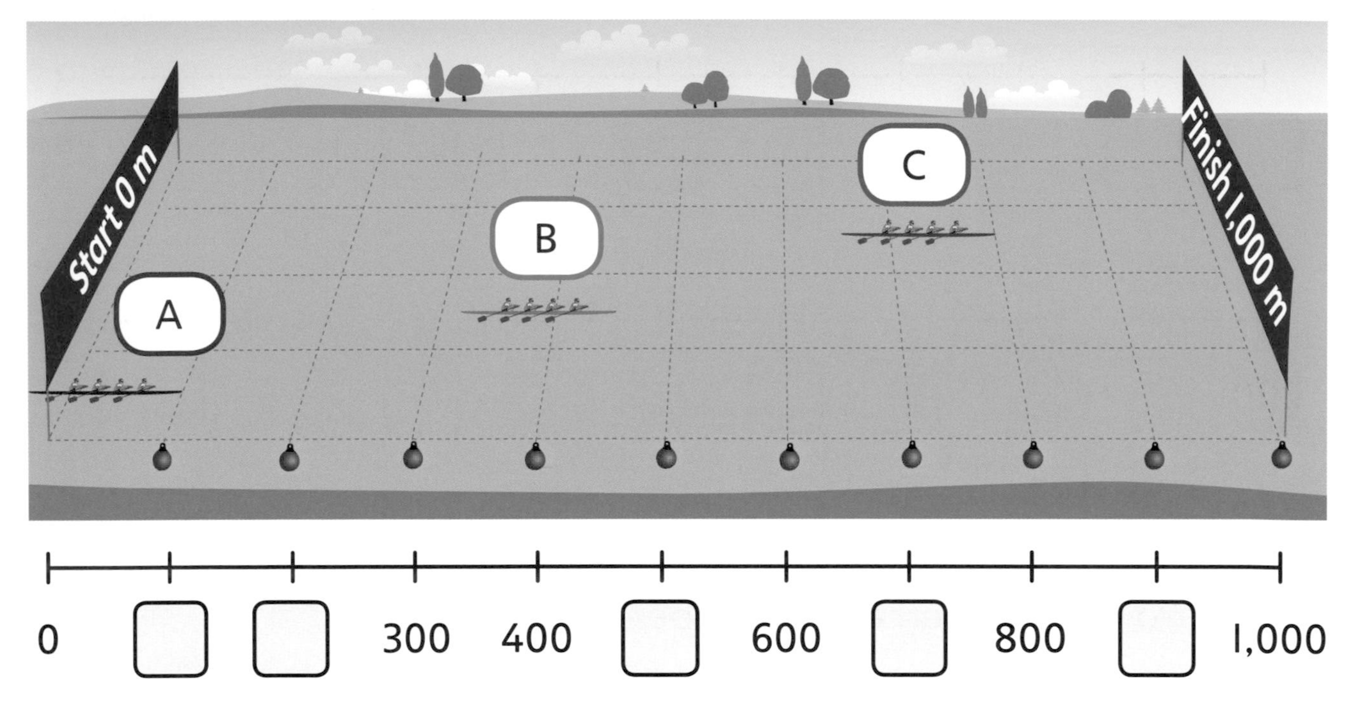

a) What are the missing numbers?

b) How far has boat A travelled?

c) How far has boat C travelled?

d) Estimate how far boat B has travelled.

2 **a)** What numbers are shown by the arrows?

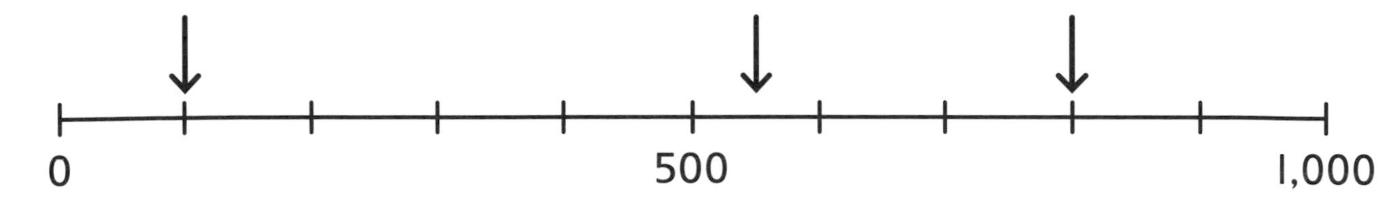

b) Point to the numbers 300, 500 and 990.

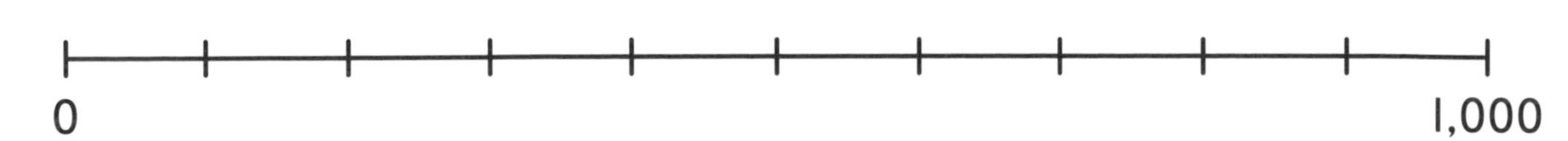

a) Work out all the missing numbers.

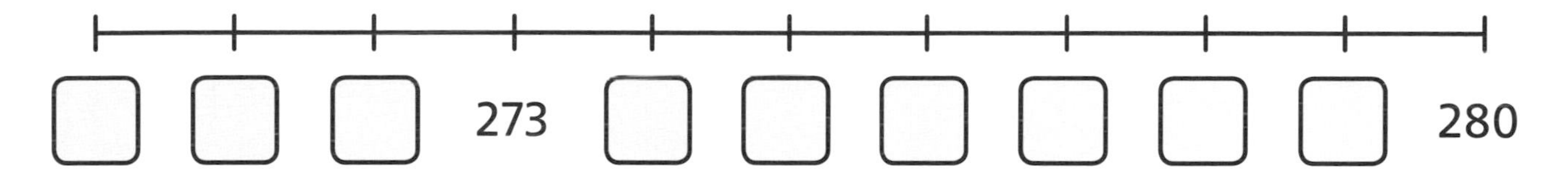

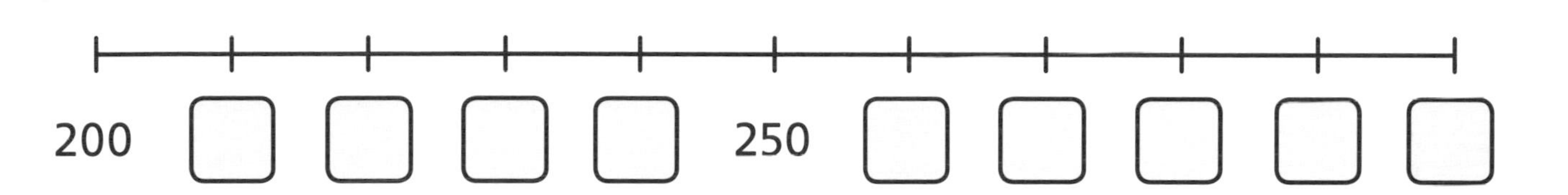

b) Point to 275 on each number line.

I don't think these number lines go up in 100s this time. I will guess and see if it works.

Last year we saw number lines that went up in 10s and 1s. I wonder if any of these do that.

→ Practice book 3A p18

The number line to 1,000 2

Discover

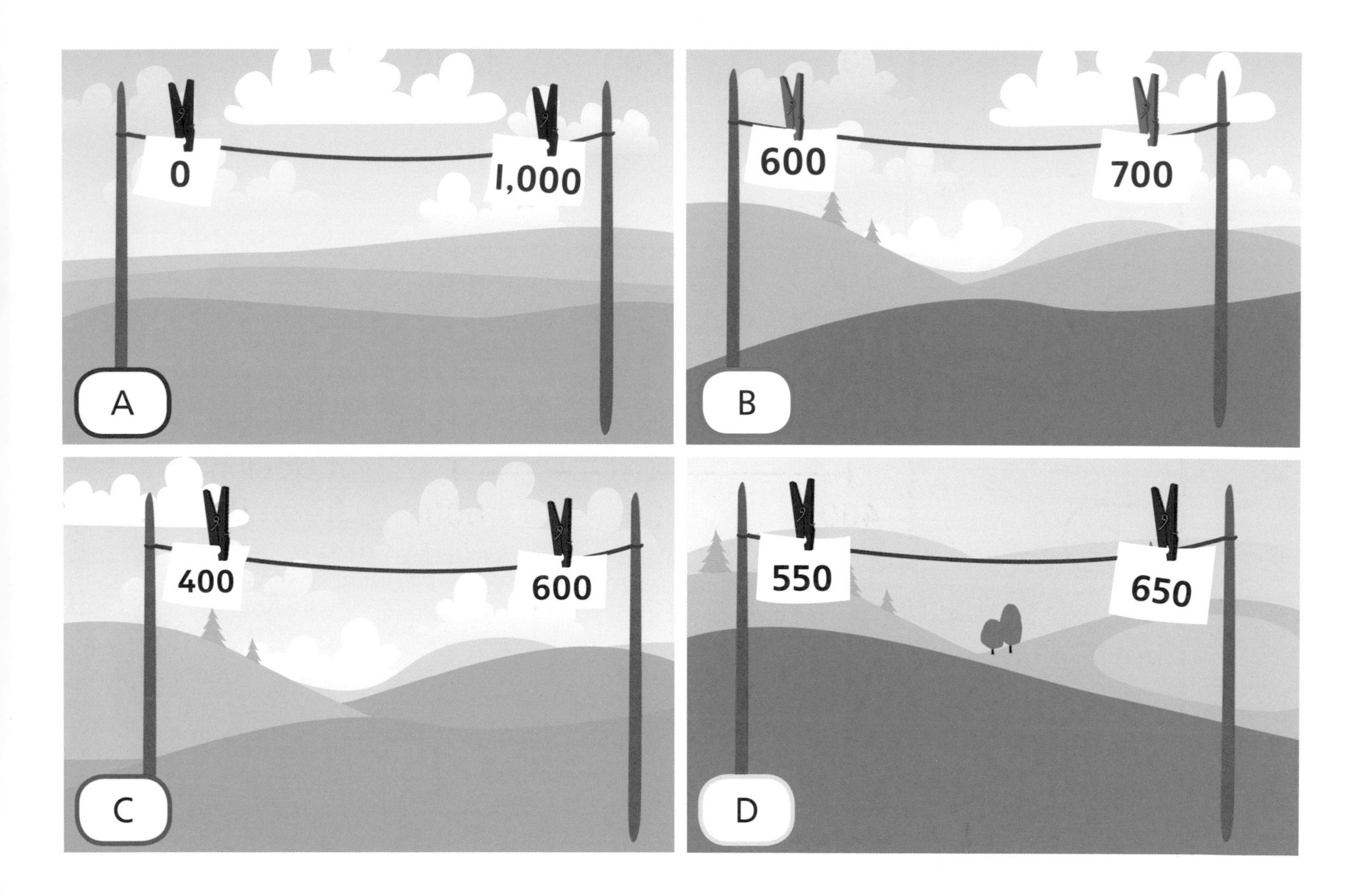

1 a) Which number lines can you peg this number card to?

500

b) Work out three numbers you can peg to line D.

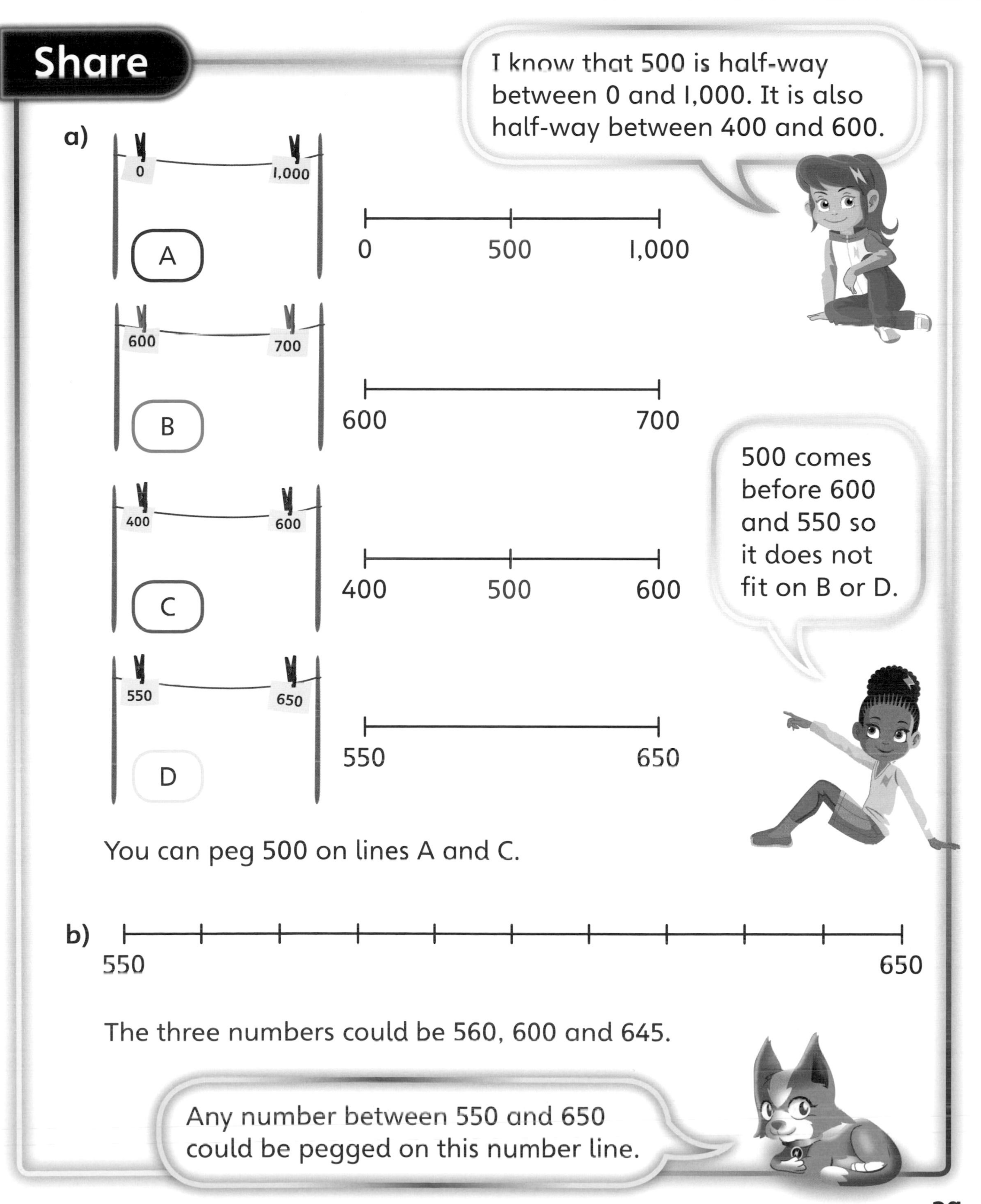
Share
I know that 500 is half-way between 0 and 1,000. It is also half-way between 400 and 600.
a)
0
1,000
A
0
500
1,000
600
700
B
600
700
500 comes before 600 and 550 so it does not fit on B or D.
400
600
C
400
500
600
550
650
D
550
650
You can peg 500 on lines A and C.
b)
550
650
The three numbers could be 560, 600 and 645.
Any number between 550 and 650 could be pegged on this number line.

Think together

1 Which of the numbers can go on these lines?

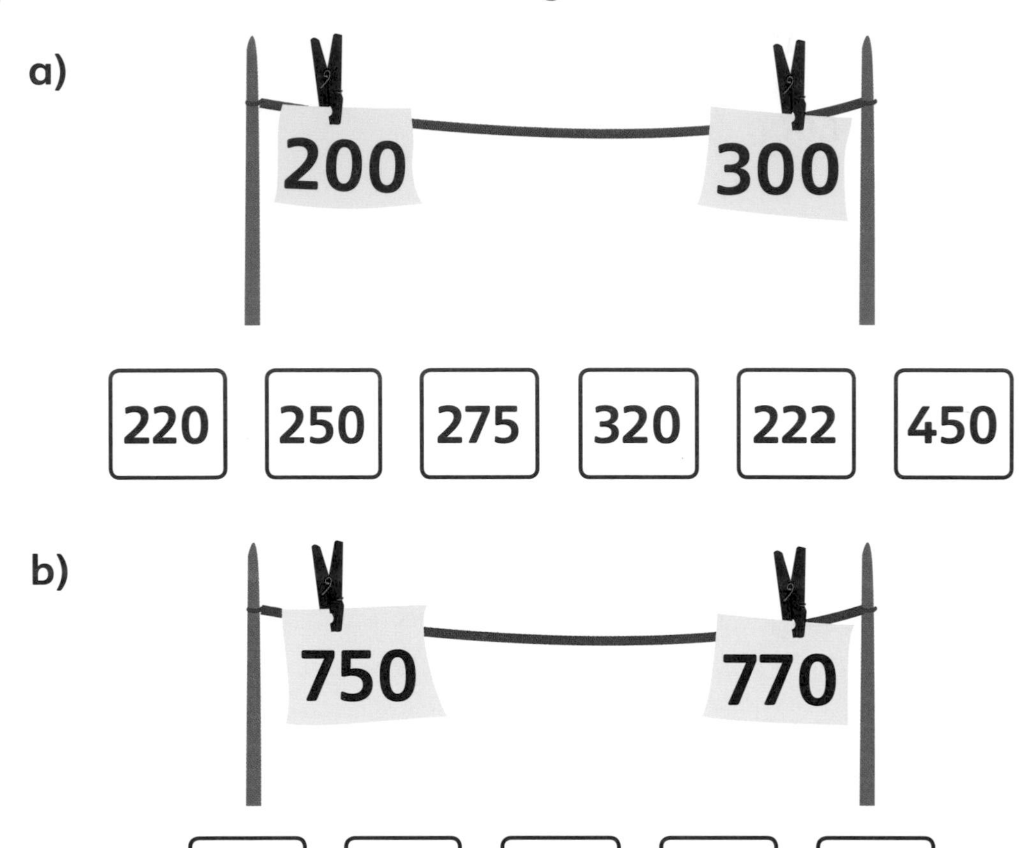

60 775 762 755 706

2 All these numbers appear on the number line.

400 ?

What could the end number be?

The end number of the number line could be ☐.

3 Is there a number that can go on every line?

A

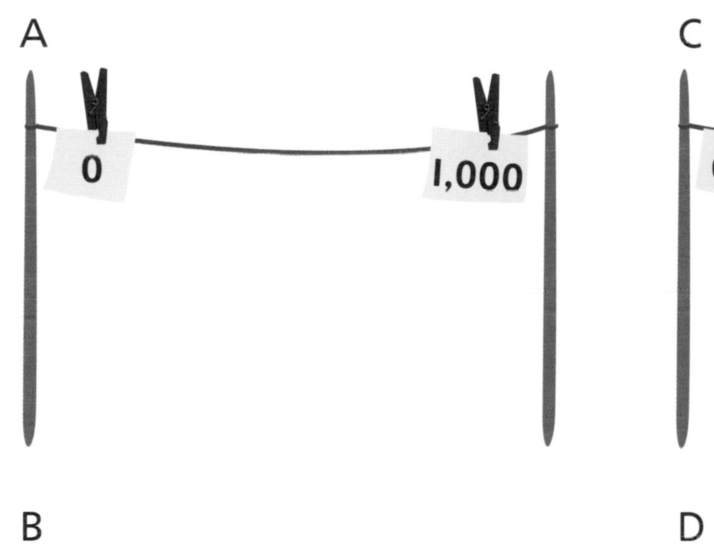

C

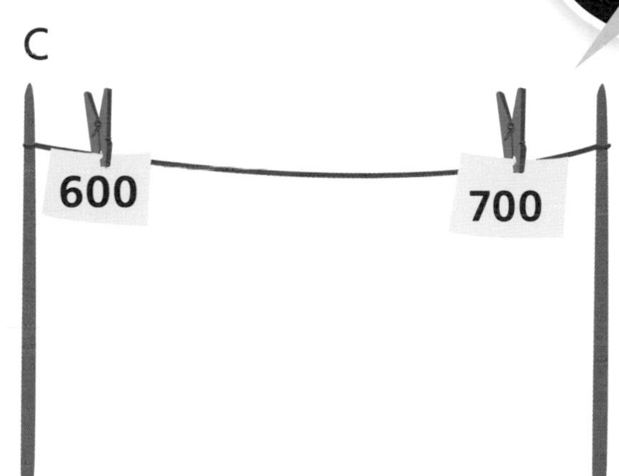

B

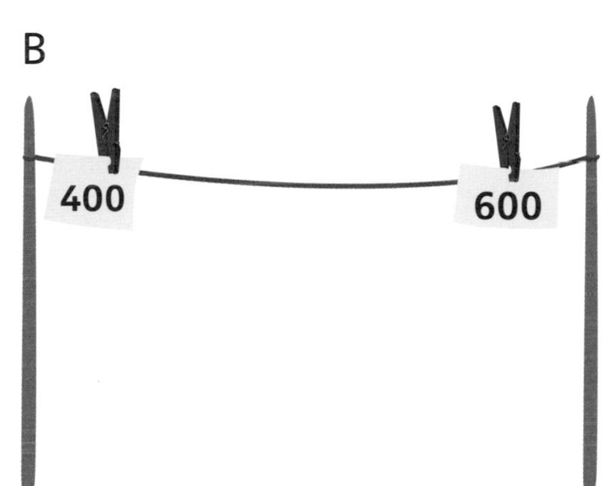

D

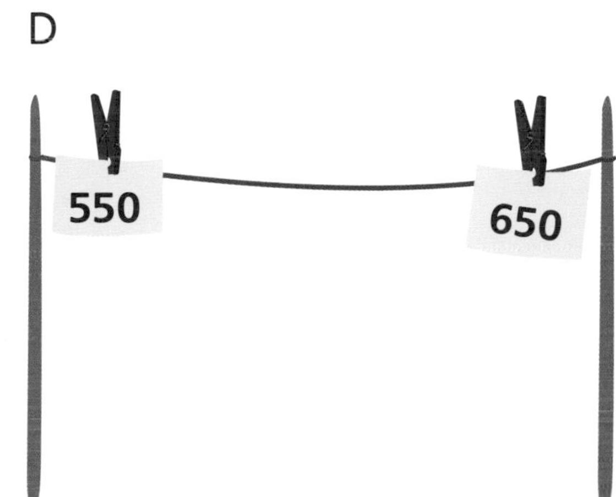

I don't think there is. Each number line has a different start and finish.

I think there might be one number that appears on every line.

→ Practice book 3A p21

Finding 1, 10 and 100 more or less

Discover

1 **a)** How many points does Amal have?

Amal receives 100 more.

How many points does he have now?

b) Holly has 204 points.

She loses 10 points.

How many points does she have now?

Share

I can represent the notes using base 10 equipment.

a)

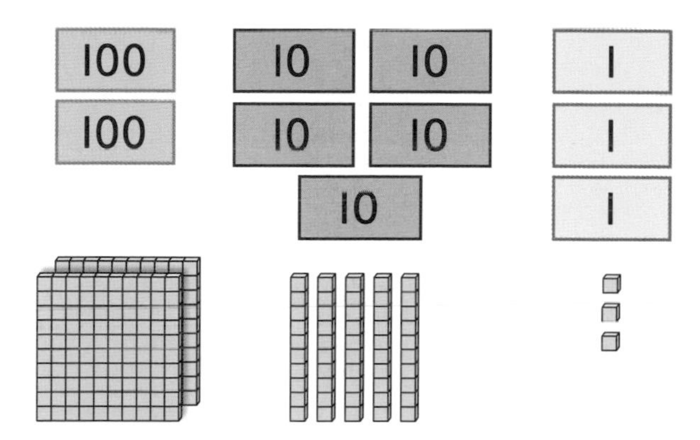
100
100
10
10
10
10
10
1
1
1

Amal has 253 points.

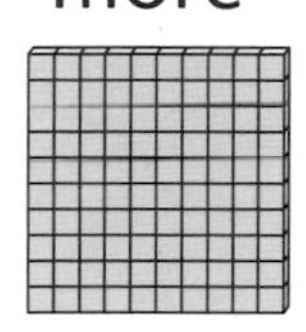
100
more

He receives 100 more.

100 more than 253 is 353.

Amal now has 353 points.

b) Holly has 204 points and loses 10.

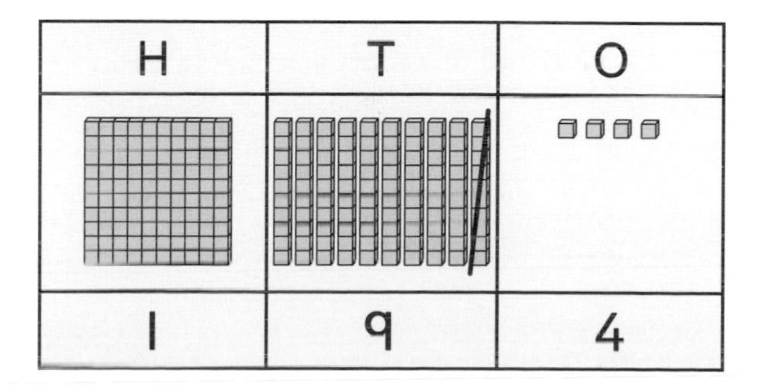

H	T	O
2	0	4

H	T	O
1	9	4

I don't have any 10s.
I need to **exchange**
1 hundred for 10 tens.

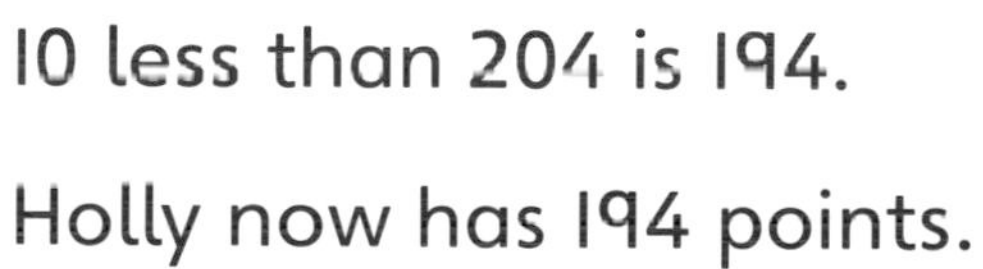

10 less than 204 is 194.

Holly now has 194 points.

Think together

1 How many points does each child have?

Child	Starting	What happens	Number of points now
Reena	100 100 10 10 1 1 100 100 10 10 1 1 100 10 10	Wins 10	
Andy		Loses 100 points	
Mo	782 points	Wins 1 point	

2 Work out these amounts.

a) 10 more than

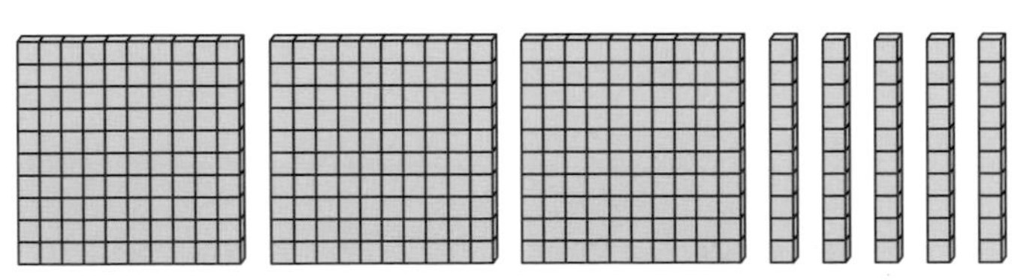

b) 100 less than

H	T	O
6	4	8

c) 1 more than 248

3 Work out 10 less than 407.

Remember, you may need to exchange.

Kate is working out 10 more than 195 .

H	T	O
1	9	5

I think the answer is 1105.

Kate

Ebo is solving the problem in the box.

The answer is 557.

Ebo

457 is 100 more than ______

What mistake has each of them made?

I can check Ebo's answer and see if the sentence is true.

I think Kate needs to exchange some 10s for 100.

→ Practice book 3A p24

Comparing numbers to 1,000 1

Discover

1 a) Are there more round lollies or square lollies?

Explain to your friend how you know.

b) Here are some triangular lollies.

Use <, > or = to compare the number of square and triangular lollies.

Write two different statements.

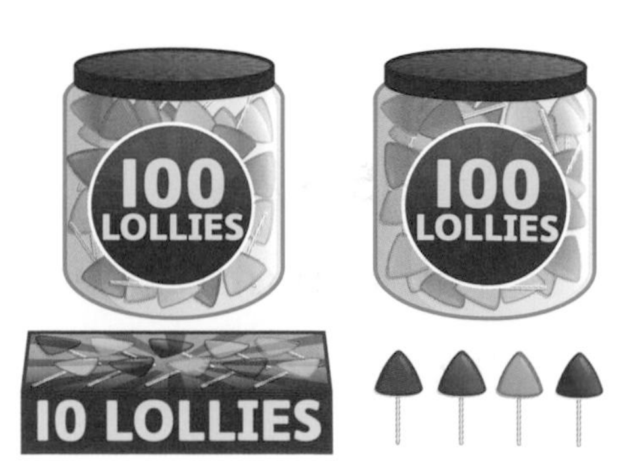

Share

a)

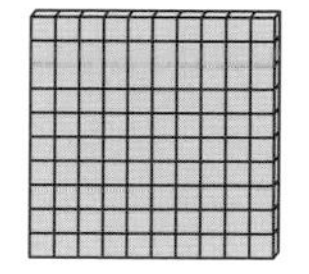

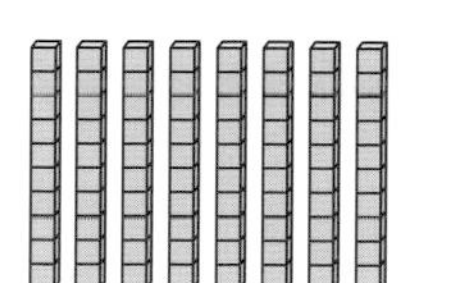

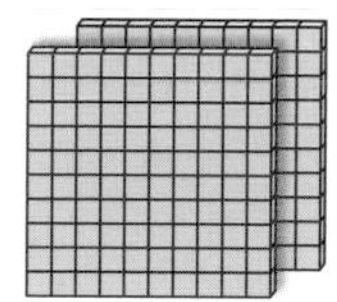

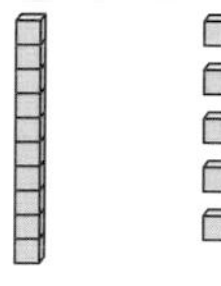

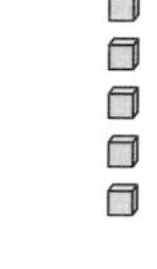

There are 183 round lollies.

There are 215 square lollies.

183 has 1 hundred,

215 has 2 hundreds.

So 215 > 183

There are more square lollies.

I compared the number of jars first. There are more jars of square lollies – so there must be more square lollies.

b) There are 215 square lollies.

There are 214 triangular lollies.

There are the same number of 100s.

There are the same number of 10s.

5 is greater than 4.

So 215 > 214

This can also be written as:

214 < 215

H	T	O
2	1	5

H	T	O
2	1	4

Think together

1 Which teacher has more books?

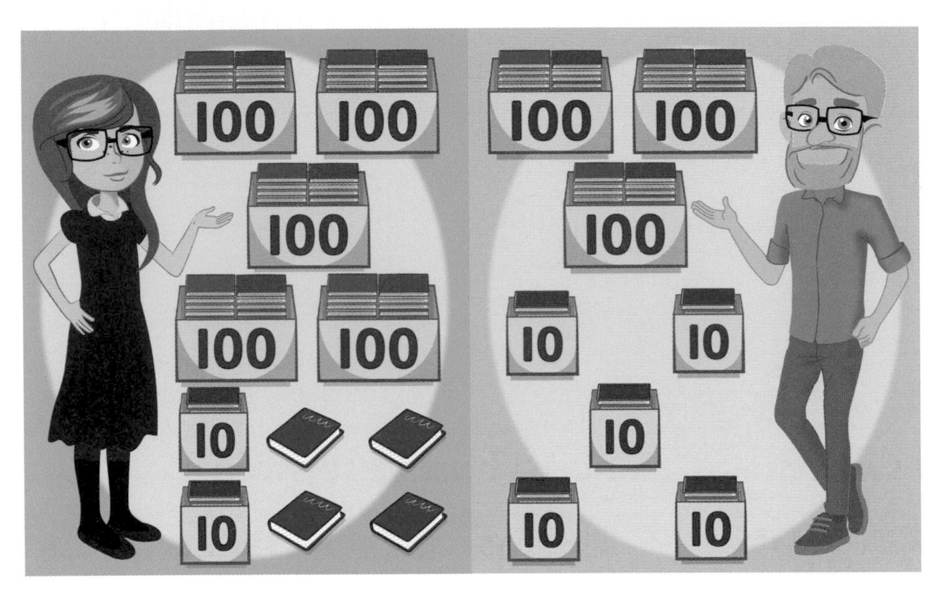

Miss Hall has ☐ books. Mr Jones has ☐ books.

☐ is greater than ☐,

so ☐ > ☐

________ has more books.

2 Complete the sentences using <, > or =

a) 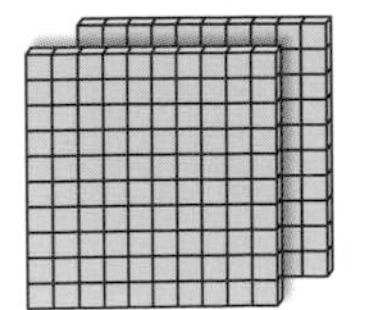  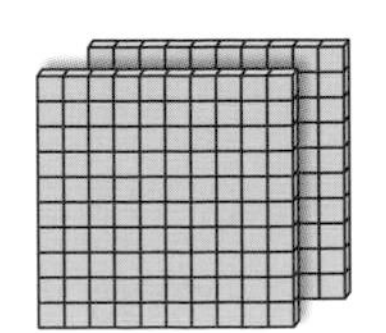 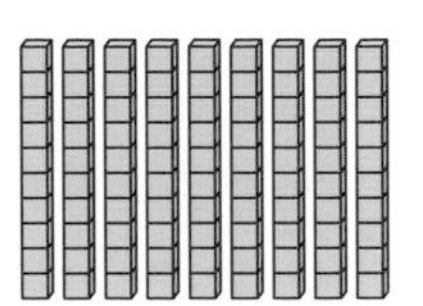

b) 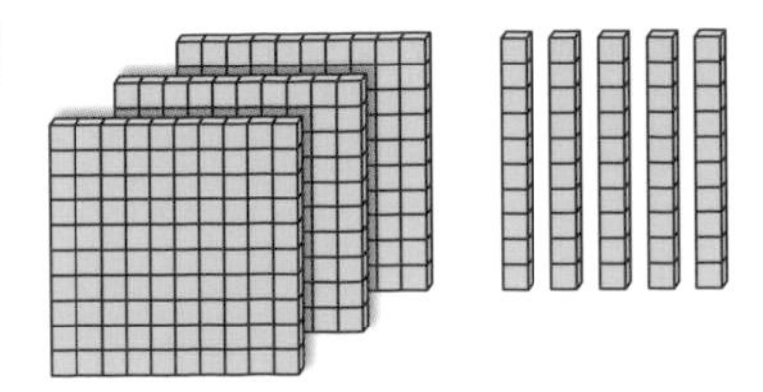 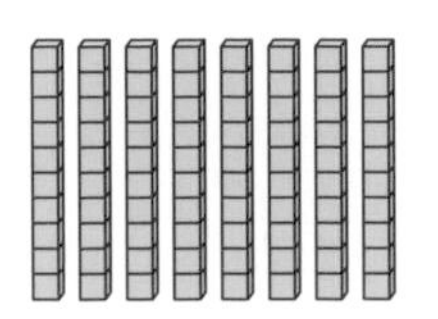

3 Complete the sentences using <, > or =

a)

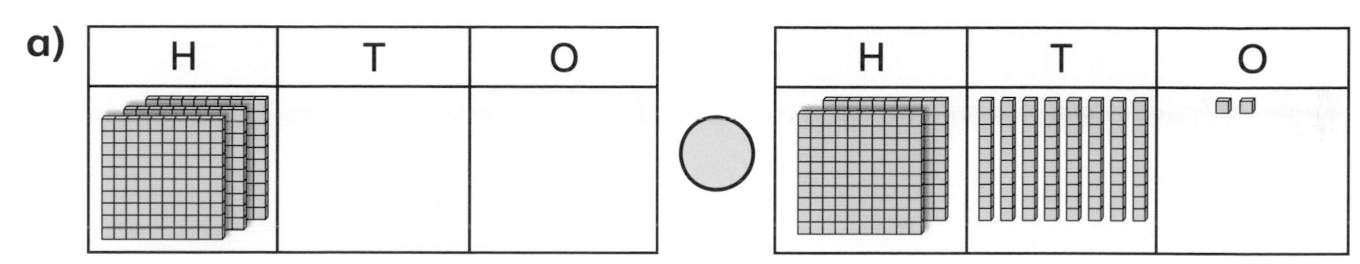

b)

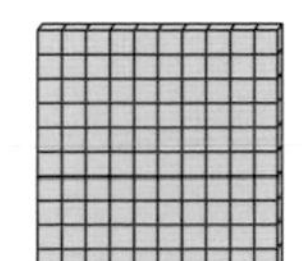
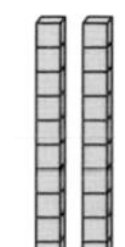

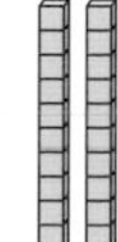
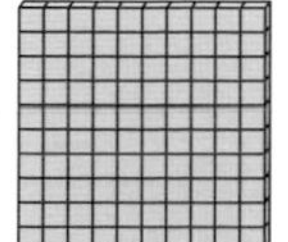

4 Danny has these sweets.

Ambika has these sweets.

Who has more sweets?

I think Ambika has more sweets as she has more jars.

I am not sure. Danny looks like he has more sweets. Look how many packs he has.

→ Practice book 3A p27

Comparing numbers to 1,000 ②

Discover

1 **a)** In each number pair shown below, which is the smaller number?

b) What could the number 3✱✱ be?

395 < 3✱✱

Share

I put each number into a place value grid. I then compared the 100s. They were the same, so I compared the 10s.

a)

H	T	O
(5)	4	2

H	T	O
(5)	8	9

H	T	O
5	(4)	2

H	T	O
5	(8)	9

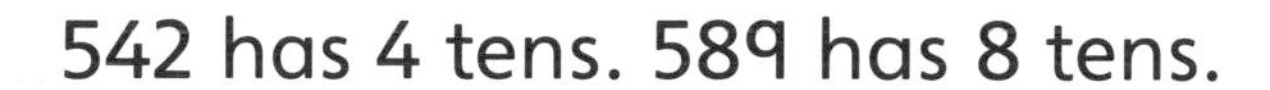

542 has 4 tens. 589 has 8 tens.

4 tens is less than 8 tens.

So 542 is less than 589, or 542 < 589.

542 is the smaller number.

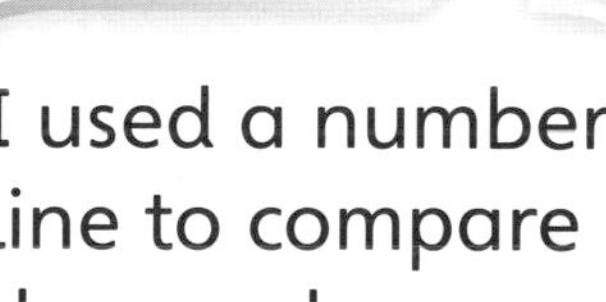

I used a number line to compare the numbers.

240 has fewer 100s than 395 so 240 is the smaller number.

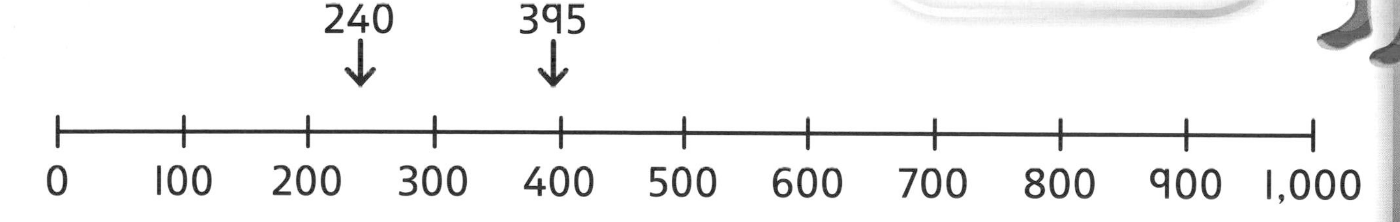

b) 395 < 3**.

The 100s are the same.

We now compare the 10s. There must be 9 tens, or 395 would be greater than 3**.

We now compare the 1s. There must be more than 5 ones.

The number could be 396, 397, 398 or 399.

Think together

1

a) Which numbers are smaller than 490?

b) Which numbers are greater than 580?

2 Complete the sentences using <, > or =

a)

H	T	O
9	4	8

H	T	O
8	2	0

b)

H	T	O
3	8	5

◯

H	T	O
3	6	8

c) 600 ◯ 950

H	T	O

H	T	O

3 Use the number line to compare these numbers.

0 ... 1,000

a) 749 ◯ 826

b) 286 ◯ 243

c) 580 ◯ 68

d) 300 ◯ 307

4 Work out the missing digits.

CHALLENGE

a)

542 is greater than 5✱6

b)

✱58 $<$ 542

c)

✱58 $<$ 395

I have found more than one answer for each one.

I wonder what the greatest digit is that can replace ✱ in ✱58.

→ Practice book 3A p30

Ordering numbers to 1,000

Discover

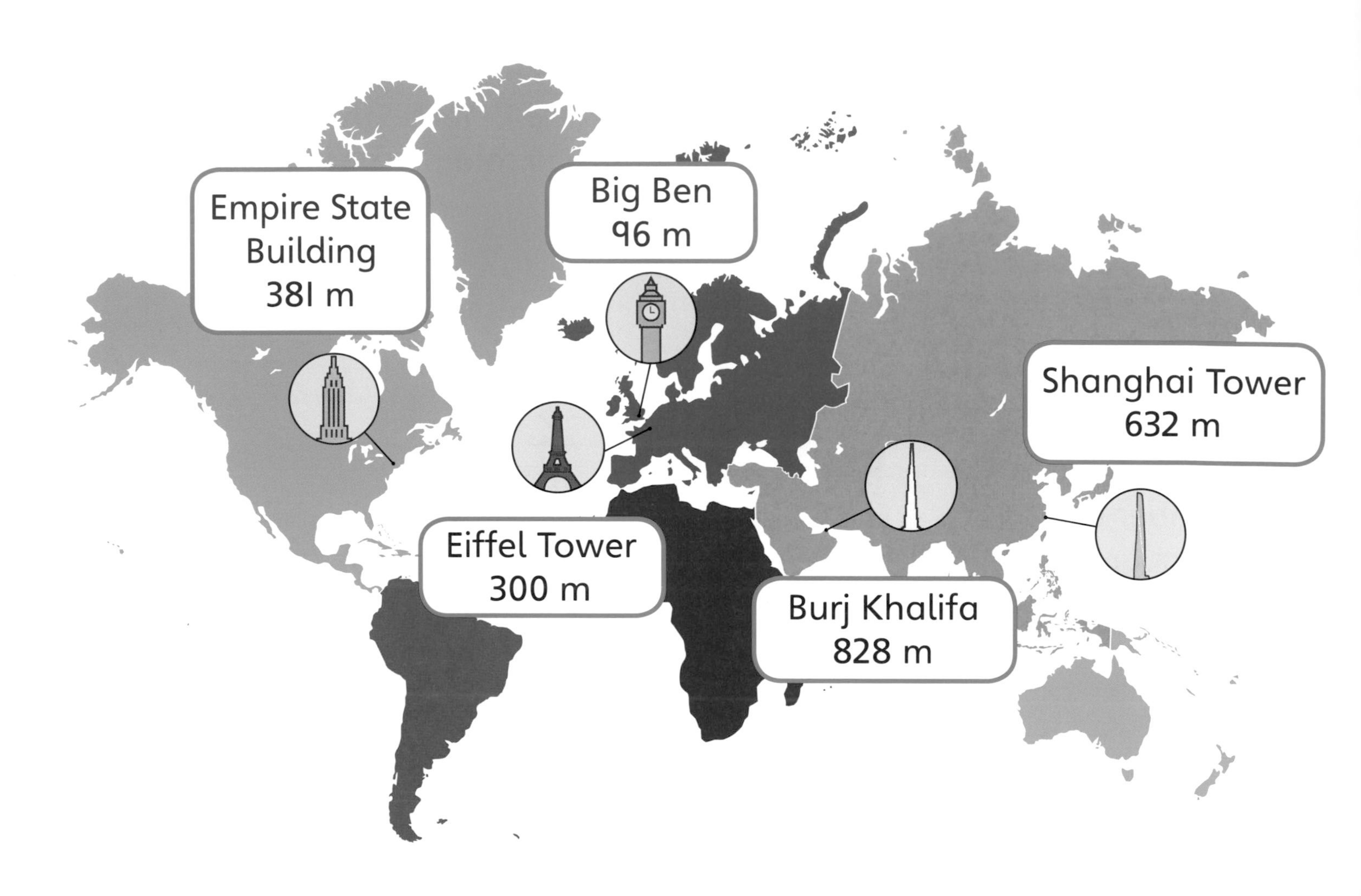

1 **a)** Which is taller, the Empire State Building or Big Ben?

b) Put the buildings in order of height.

Start with the shortest.

Share

a) Empire State Building

H	T	O
3	8	1

Big Ben

H	T	O
	9	6

The digit in the hundreds column for the Empire State Building is greater than the digit in the hundreds column for Big Ben.

So 381 > 96

The Empire State Building is taller than Big Ben.

I put the numbers onto a number line to compare.

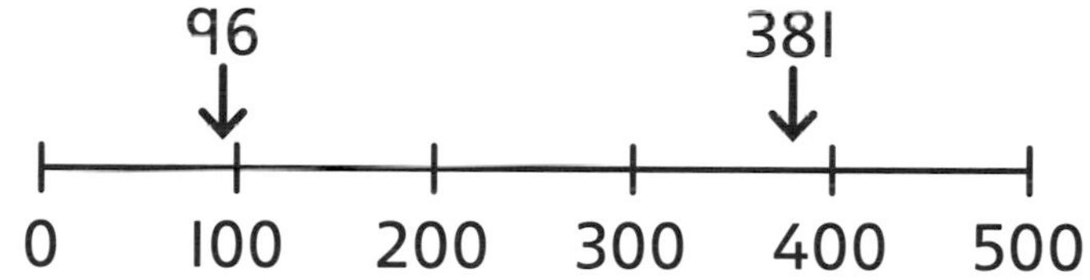

b)

I compared the 100s, then the 10s, then the 1s.

	H	T	O
Big Ben		9	6
Eiffel Tower	3	0	0
Empire State Building	3	8	1
Shanghai Tower	6	3	2
Burj Khalifa	8	2	8

Think together

1 Put these ships in order of length. Start with the shortest.

Cruise ship 238 m	Ferry 82 m	Container ship 285 m

	H	T	O
Cruise ship			
Ferry			
Container ship			

________________ , ________________ , ________________ .

shortest longest

2 Four boxes contain some counters.

Put the numbers of counters in order.

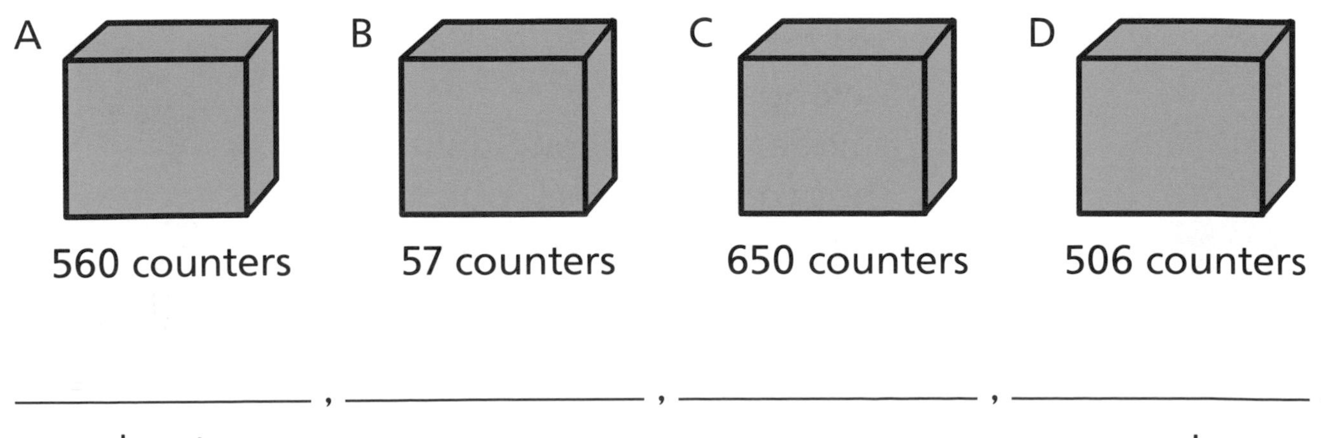

________________ , ________________ , ________________ , ________________ .

least most

3 Are these numbers in an order?

740, 704, 470, 407, 74

If they are, what order are they in?

If not, can you put them in order?

Use the number line to help you.

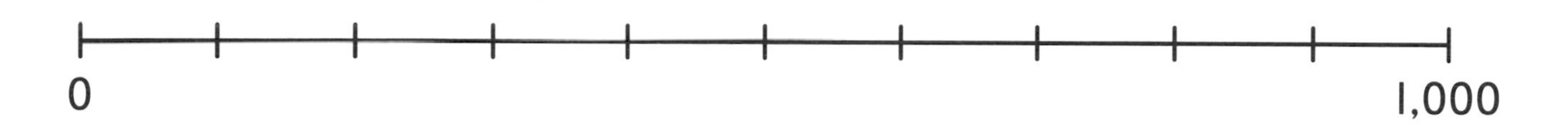

4 These numbers are in order from smallest to greatest .

Find the missing digits.

I think there is more than one answer.

I wonder what is the smallest and largest digit that can go in each box.

→ Practice book 3A p33

Counting in 50s

Discover

1. a) How many stars on each flag?

   How many stars on 4 flags?

   b) Sylvie counts 350 stars.

   How many flags has she counted?

Share

a)

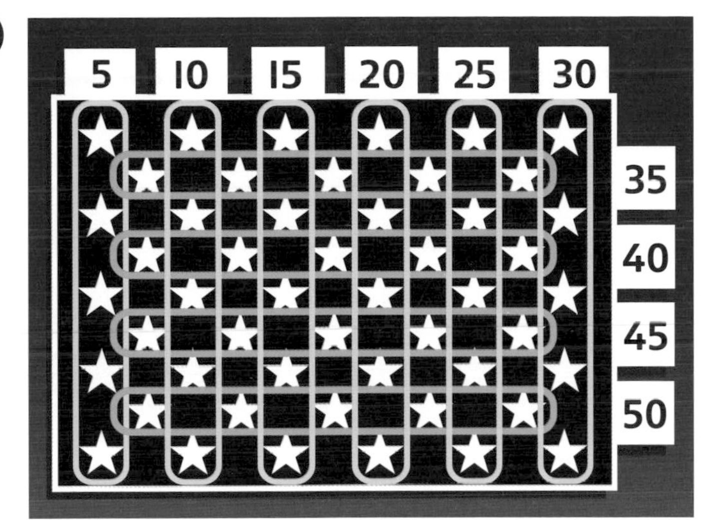

I counted the stars in 5s. I saw that they were in columns and rows.

There are 50 stars on each flag.

I counted up in 50s.

There are 200 stars on 4 flags.

b)

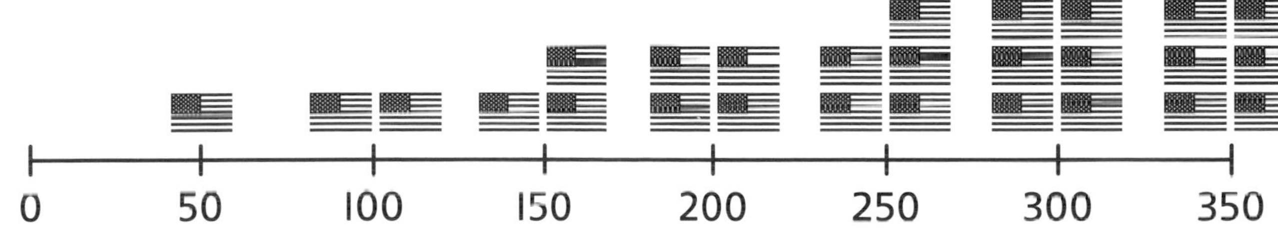

Sylvie counted 350 stars on 7 flags.

Think together

1 Work out the missing numbers.

Number of flags	0	1	2	3	4	5	6	7	8	9	10
Number of stars	0	50	100								

2 a) How many stars on 15 flags?

There are ☐ stars on 15 flags.

b) How many stars on 16 flags?

There are ☐ stars on 16 flags.

c) How many stars on 17 flags?

There are ☐ stars on 17 flags.

I wonder if I have to start at 0 each time.

3 Ben counts 1,000 stars. How many flags are there?

There are 1,000 stars on ☐ flags.

4 Here is a 0 to 1,000 number line.

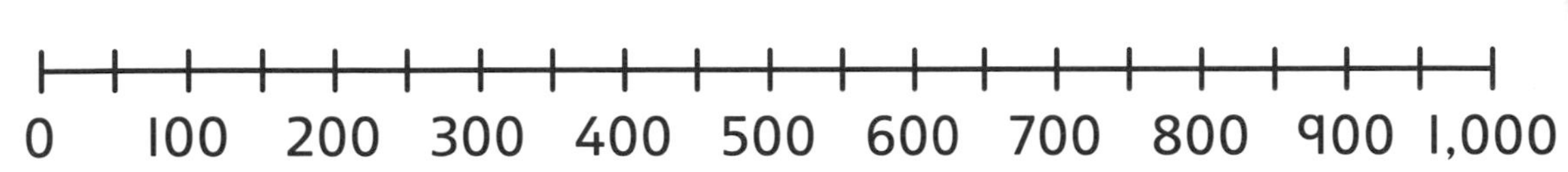

I can use the number line to count in 50s from 0 to 1,000 and back again.
Kate

Which of these numbers will Kate say?

95	750	650	400	50	505	355	250

What did you notice about all the numbers Kate says?

Think of a rule about the numbers Kate will say.

The number line goes up in 100s. I am not sure how Kate can use it to count in 50s.

I have noticed something interesting about the last two digits of the numbers Kate says.

→ Practice book 3A p36

End of unit check

1 What number is shown?

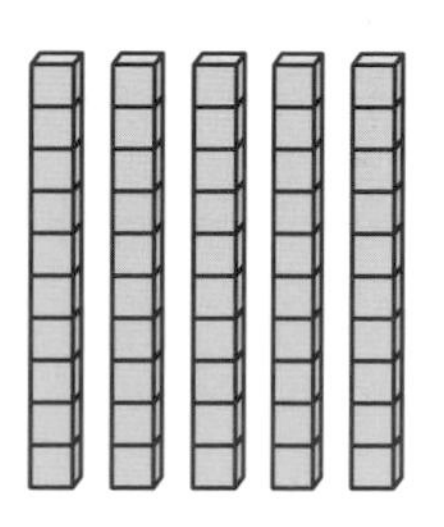
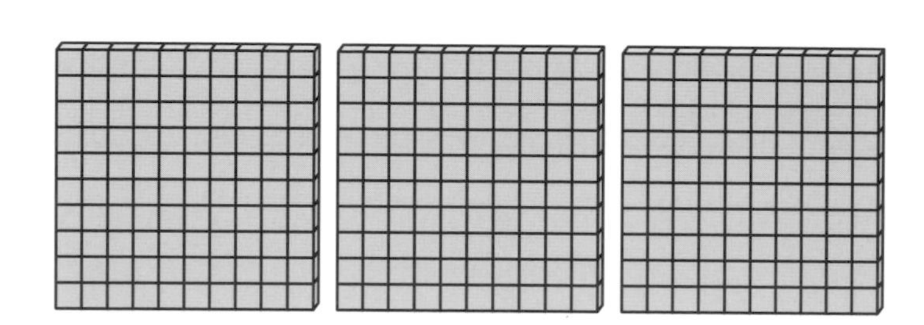

A 325 B 352 C 523 D 532

2 Which number line shows the arrow pointing to 350?

A
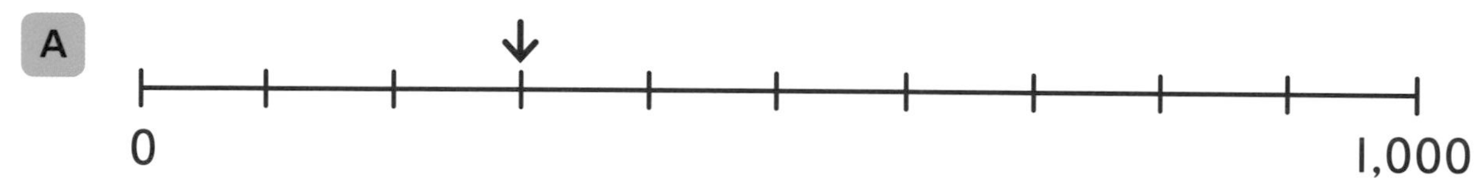

B
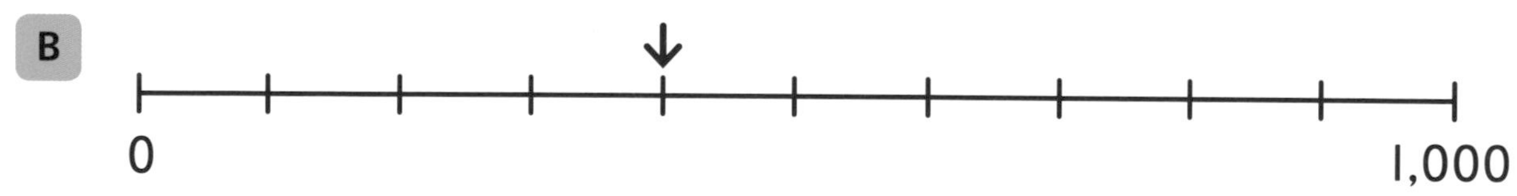

C
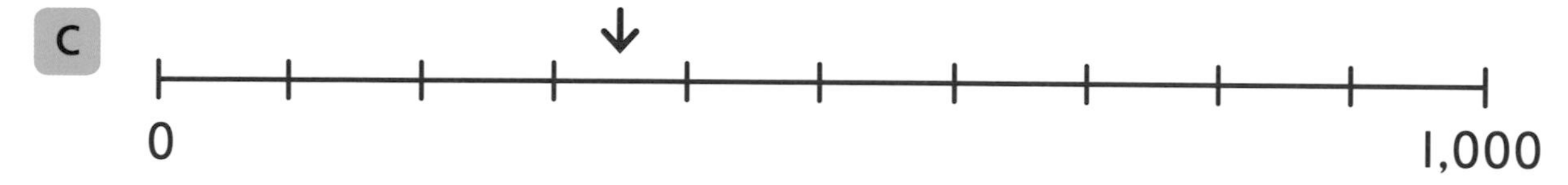

D
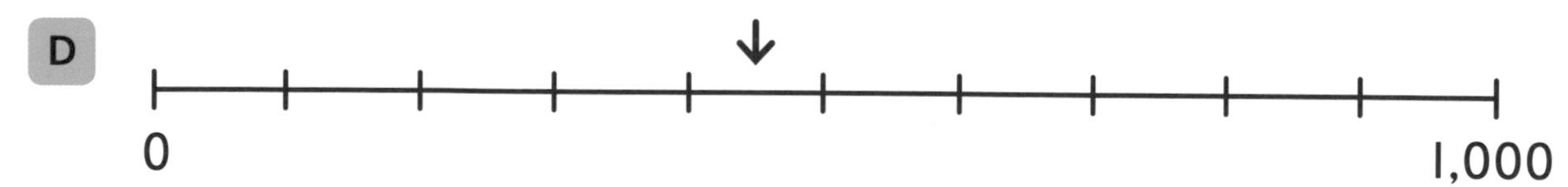

3 10 more than ☐ is 385.

A 485 B 285 C 395 D 375

4 Which statement is correct?

H	T	O
4	2	9

H	T	O
3	8	1

A 429 < 381

B 429 = 381

C 429 > 381

D None of them

5 The number track goes up in 50s.

200	250	300				500

What number should be in the shaded box?

A 350 B 303 C 450 D 499

6 Which set of numbers is in order from smallest to greatest?

A 54, 540, 504, 450

B 450, 504, 54, 540

C 540, 504, 450, 54

D 54, 450, 504, 540

7 There are three boxes of counters.

Box X contains 160 counters. Box Y contains 84 counters.
Box Z contains 100 more counters than Box Y.

Put the boxes in order. Start with the one with the fewest counters.

→ Practice book 3A p39

Unit 2
Addition and subtraction 1

In this unit we will ...

- Add 1s and 10s to 3-digit numbers
- Subtract 1s and 10s from 3-digit numbers
- Add and subtract 3-digit and 2-digit numbers
- Learn when to exchange 1s, 10s and 100s
- Add and subtract using mental and written methods

Do you remember how to use place value? What numbers do these represent?

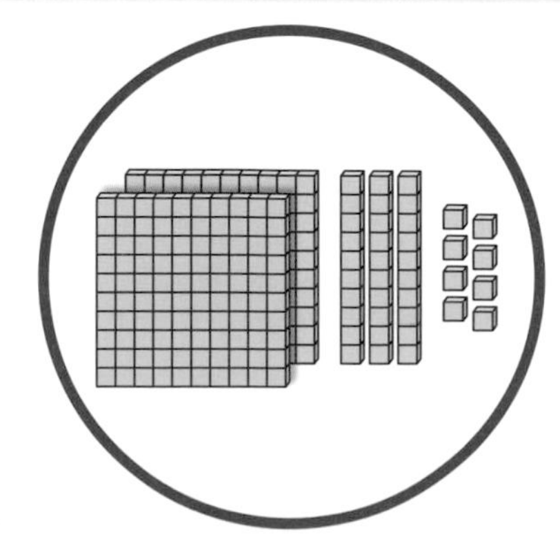

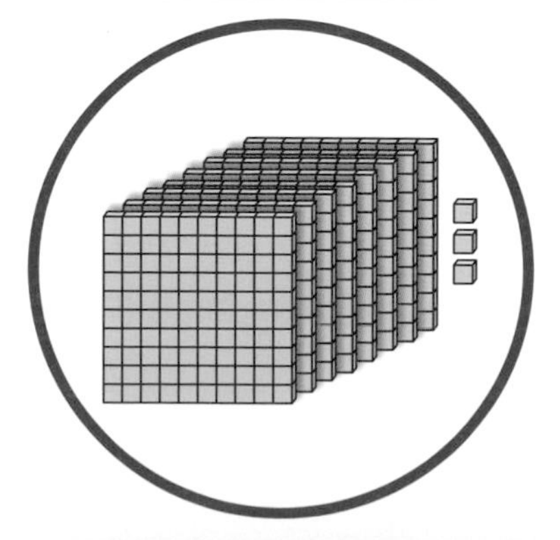

We will need some maths words.
Are any of these new?

addition **subtraction** **mental method**

column method **exchange**

We need this too! Use it to write the number two hundred and thirty-four using digits.

H	T	O

Adding and subtracting 100s

Discover

1 a) The lorry delivers 2 more packs of bricks. Each pack holds 100 bricks.

How many bricks does Amal now have in total?

b) How many bricks are left on the lorry?

Share

a) There are 100 bricks in each pack and 2 more packs were delivered.

I will count in 100s.

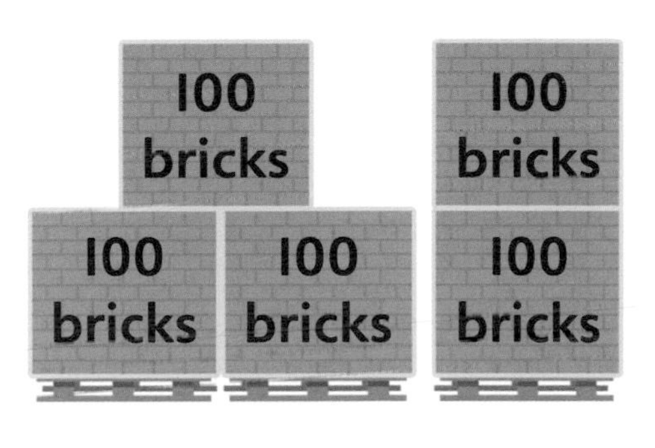

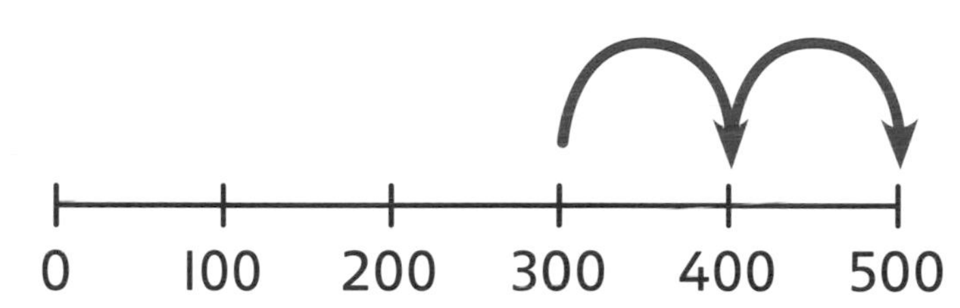

I will use number bonds to add the 100s.

I know that 3 + 2 = 5. So, there are 5 hundreds in total.

There are 3 hundreds and 2 hundreds.

300 + 200 = 500

Amal has 500 bricks now.

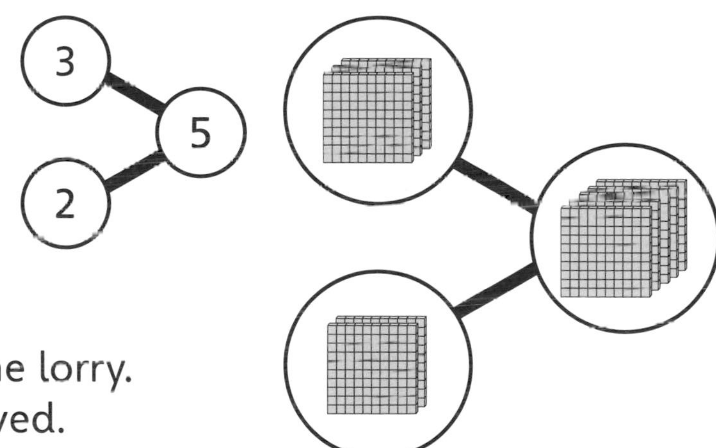

b) There were 4 hundreds on the lorry. Then 2 hundreds were removed.

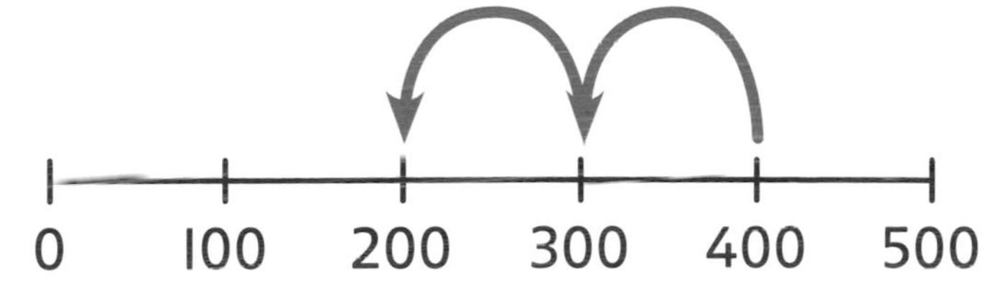

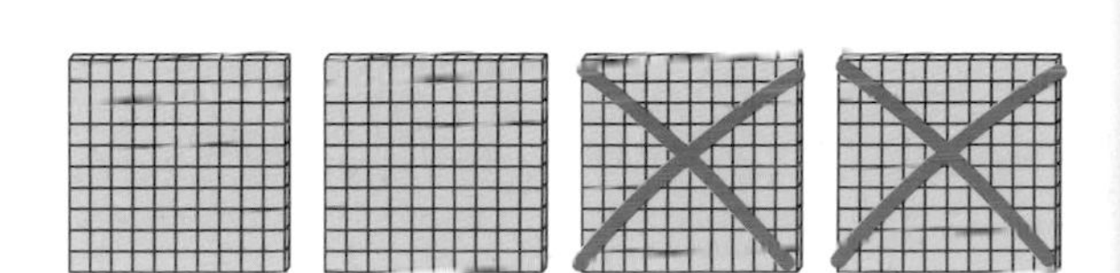

400 – 200 = 200

200 bricks are left on the lorry.

Think together

1 There are 7 boxes. There are 100 hinges in each box.

3 boxes have been used up. How many hinges are left?

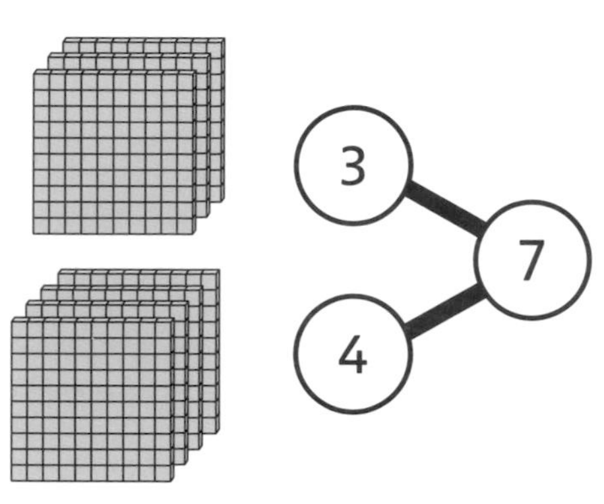

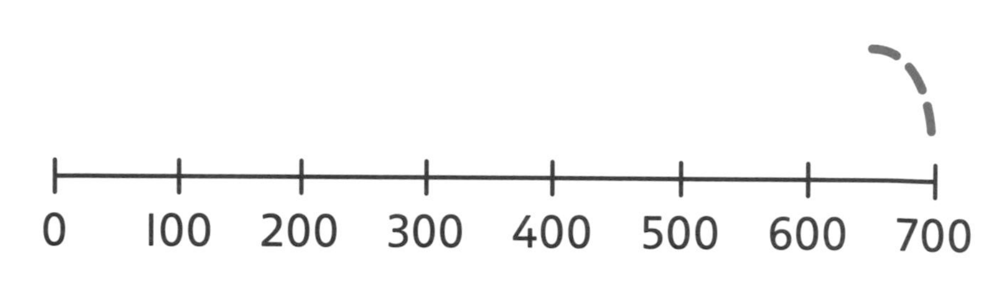

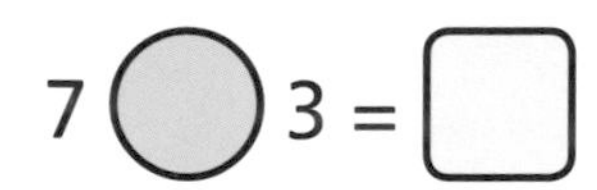

7 ◯ 3 = ☐

☐ ◯ 300 = ☐

There are ☐ hundreds left.

There are ☐ hinges left.

2 **a)** A builder uses 400 nails.
How many does she have left?

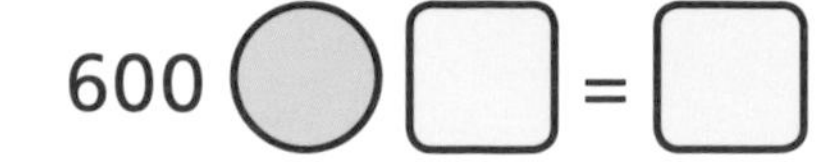

600 ◯ ☐ = ☐

She has ☐ nails left.

b) How many screws does the builder have in total?

☐ hundreds ◯ ☐ hundreds

= ☐ hundreds

He has ☐ screws in total.

3 Explain the mistake.

There are 100 bolts in each box.

The builders need 600 bolts in total.

How many more boxes do they need?

3 + 6 = 9
They need 900 boxes.
That can't be right,
I'll try again!

4 What other additions and subtractions can you find using this fact?

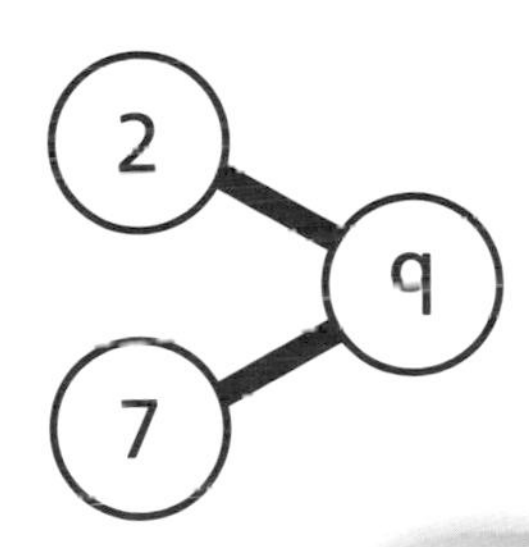

I will think of fact families. I will try to find 8 facts using 100s.

I will try adding 100s.

→ Practice book 3A p42

Adding and subtracting a 3-digit number and 1s

Discover

1 a) Once the people outside enter, how many visitors will be in the museum in total?

b) One person then leaves.

How many people are left in the museum?

Share

a) There are 245 people in the museum. 4 more arrive.

We need to work out 245 + 4 = ☐

Counting on in 1s works, but it is easy to make a mistake.

H	T	O
2	4	9

Use number bonds to add the 1s.

5 + 4 = 9

245 + 4 = 249

There will be 249 visitors in the museum in total.

b) One less than 249 is 248.

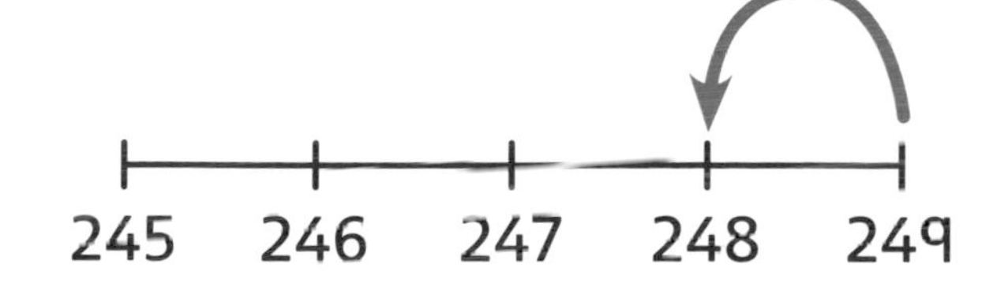

248 people are left in the museum.

Think together

1 At 12 o'clock there were 319 people in the museum. Then 7 left.

How many people are now in the museum?

H	T	O
3	1	9

I will work out the number bonds within 10 to solve this.

9 − ☐ = ☐

Now there are ☐ ones.

319 − ☐ = ☐

☐ people are now in the museum.

2 At 1 pm there were 291 people in the museum. 6 more people arrived. How many are there now?

1 ◯ 6 = ☐

291 ◯ ☐ = ☐

Now there are ☐ people.

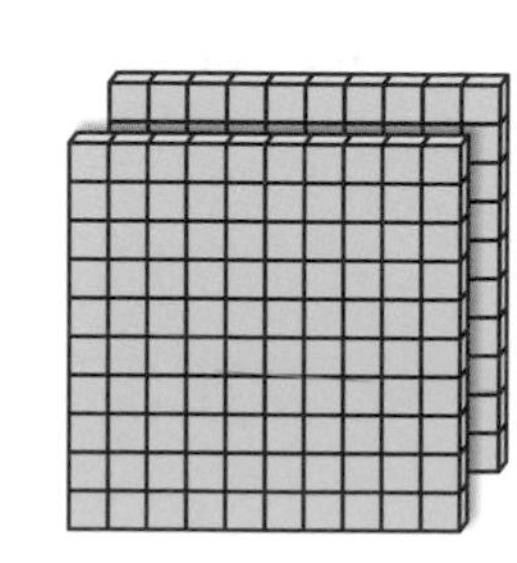

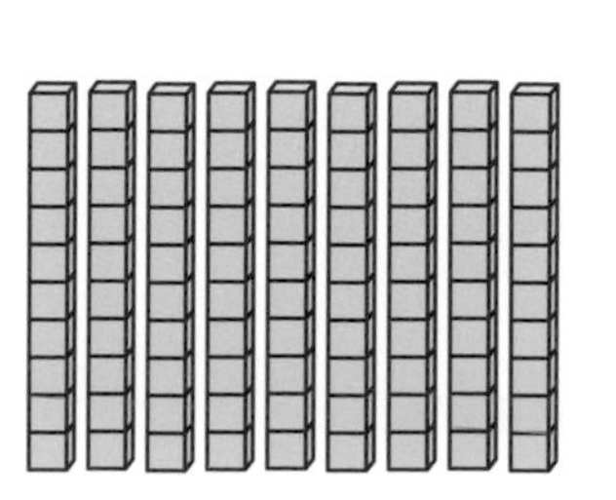

3 Work out the missing information.

Time	People in museum	Then	Now there are
2 pm	204	3 arrive	
3 pm	198	8 leave	
4 pm	158		152 people
5 pm		4 leave	115 people

4 How many solutions can you find for each calculation?

0 1 2 3 4 5 6 7 8 9

a) 4 3 ☐ + ☐ = 4 3 5

b) 4 3 ☐ – ☐ = 4 3 5

I think I found all the solutions.

I wonder how I can tell if I have found them all.

→ Practice book 3A p45

Adding a 3-digit number and 1s

Discover

1 **a)** Solve the additions.

b) Use the cards to complete this addition.

1 3 5 7

☐☐☐ + ☐ = 160

Share

a) 571 + 3 can be solved like last lesson but 135 + 7 is different.

571 + 3 = ?
I can add the 1s.
1 + 3 = 4

135 + 7 = ?
I will add the 1s.
5 + 7 = 12
Is the answer 1,312?

571 + 3 = 574

135 + 7 = ?
We need 5 more to make the next 10.
Then add on the 2.

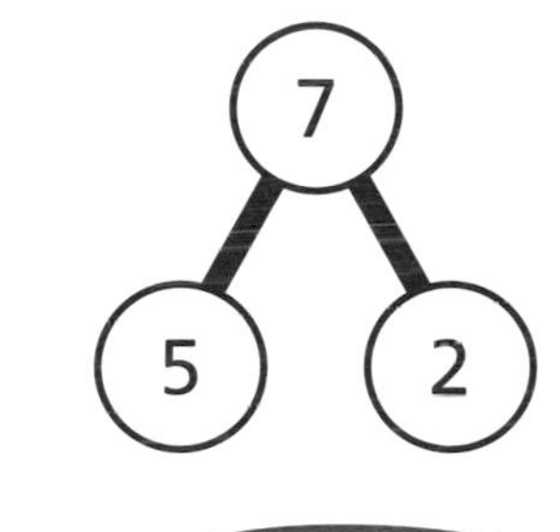

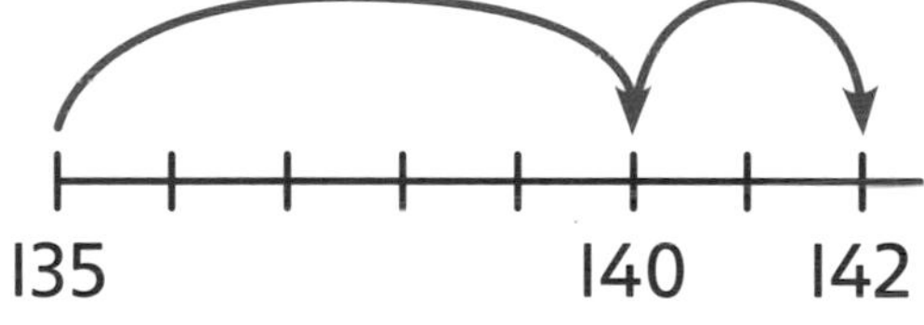

The 5 ones and 5 more make a ten.
Exchange 10 ones for 1 ten.

H	T	O

H	T	O

135 + 7 = 1 hundred + 4 tens + 2 ones = 142

b) 153 + 7 = 160 and 157 + 3 = 160

These both total 160, because 7 + 3 = 10.

Think together

1 Solve this addition.

3 | 1 | 6 | + | 5

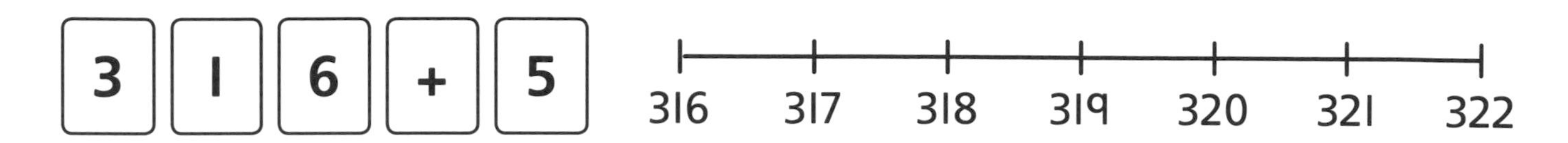

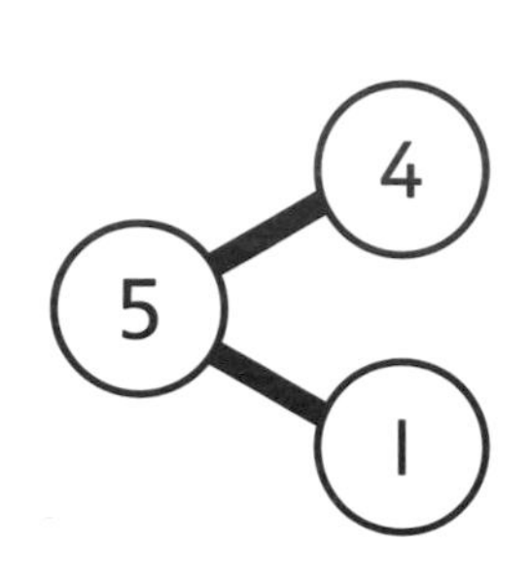

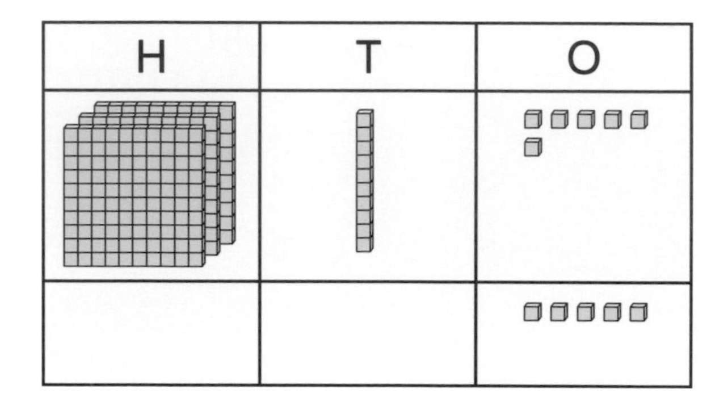

6 ones + 5 ones = ☐ ones

☐ ones = 1 ten and ☐ ones

316 + 5 = ☐ hundreds + ☐ tens + ☐ ones = ☐

2 Which addition changes the 10s digit? Can you tell before finding the answers?

A 2 | 4 | 8 | + | 6

B 8 | 4 | 2 | + | 6

Now solve the additions.

a) 8 ones + 6 ones = ☐ ones

☐ ones = 1 ten + ☐ ones

248 + 6 = ☐

b) 2 ones + 6 ones = ☐ ones

842 + 6 = ☐

3 

Find more than 4 different solutions. Add a number less than 10.

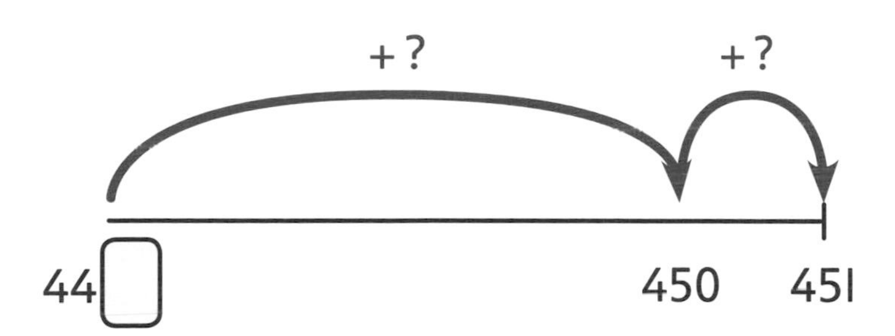

I want to find all the solutions.

4 Find an example to support Astrid's idea.

When I add a 3-digit number and 1s, the 10s digit increases by 1.

Is it always true?

Can the 10s digit ever increase by 2?

Discuss with a partner and share your reasons with the class.

→ Practice book 3A p48

Subtracting 1s from a 3-digit number

Discover

1 **a)** How many parcels do they still have to deliver?

b) Show the subtraction on a number line.

Share

a) There were 151 parcels. Then 6 were delivered.

151 − 6 = ☐

Can I subtract the 1s?
I could do 6 − 1 = 5.
So is the answer 155?

The answer can't be 155, because that would mean they have more parcels than they started with.

Exchange 1 ten for 10 ones.

H	T	O

H	T	O

They still have 145 parcels to deliver.

b) First jump back 1 to 150 and then another 5 to 145.

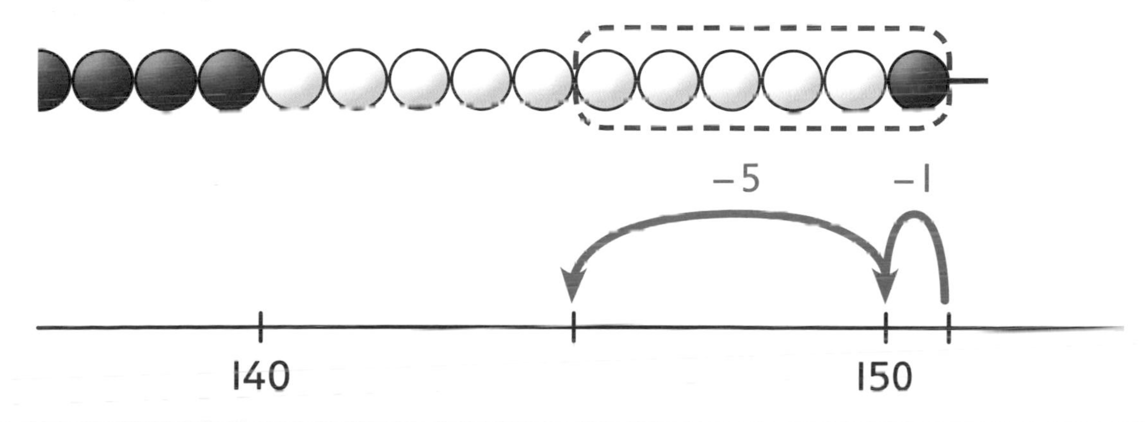

Think together

1 They have 145 parcels, then deliver 8.

How many are left?

H	T	O

H	T	O

15 ones – 8 ones = ☐ ones

145 – 8 = ☐ hundred + ☐ tens + ☐ ones

145 – 8 = ☐

2 There are 244 parcels in a green van. 9 are delivered.

There are 239 parcels in a blue van. 3 are delivered.

Which van has more parcels left?

244 – 9 = ☐

239 – ☐ = ☐

244 – 9 ◯ 239 – 3

The ______________ van has more parcels left.

I wonder if both subtractions need an exchange?

3 There are some parcels in a van. 6 are delivered. Now there are 205 parcels.

How many parcels did they start with?

205 = ☐ – 6

4 Dexter has these two subtractions:

250 – 7 = ☐ and 205 – 7 = ☐.

How can he solve them?

I am not sure what to do when there is a 0 in the tens or the ones column.

I think sometimes it is better to use a number line to jump back.

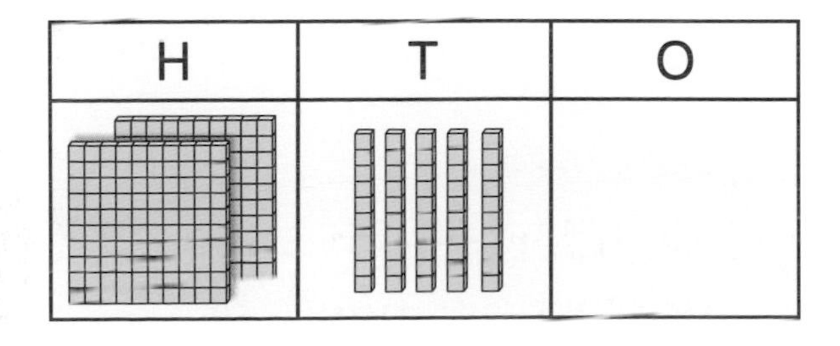

H	T	O

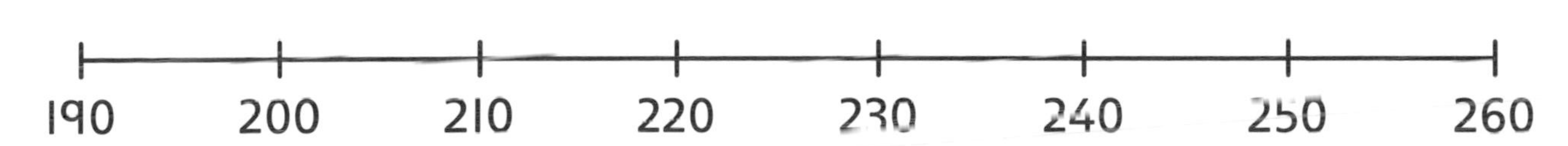

→ Practice book 3A p51

Adding and subtracting a 3-digit number and 10s

Discover

1. a) Aki has made 351. He adds 3 beads to the 10s pole.

   Write this as an addition and find the total shown.

   b) Reena then takes 1 bead from the 10s pole.

   What number does the abacus show now?

Share

a) Aki adds 3 beads to the 10s pole. So he is adding 30.

351 + 30 = ?

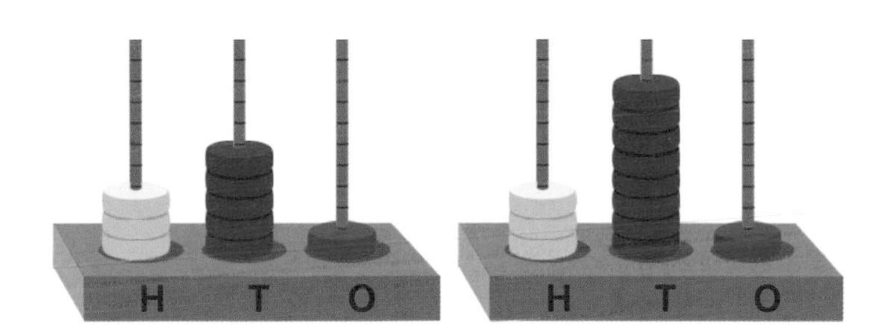

H	T	O

I know 5 + 3 = 8, and I can use this to work out the 10s.

Now there are 8 tens.

50 + 30 = 80

351 + 30 = 381

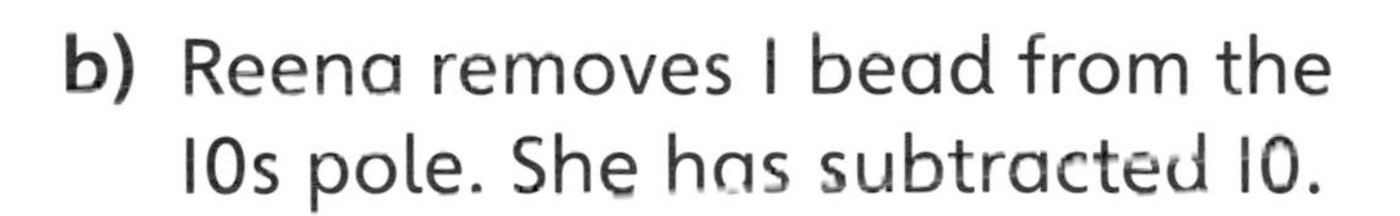

b) Reena removes 1 bead from the 10s pole. She has subtracted 10.

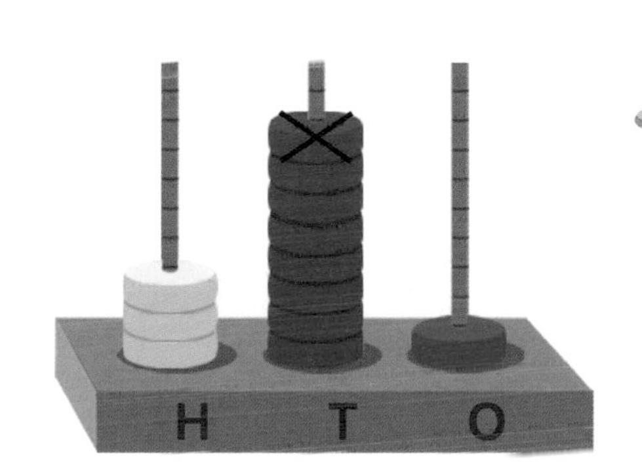

8 – 1 = 7

80 – 10 = 70

381 – 10 = 371

The 10s digit has decreased by 1.

The abacus now shows 371.

H	T	O

Think together

1 Ana takes 5 beads from the 10s pole. Show this as a subtraction.

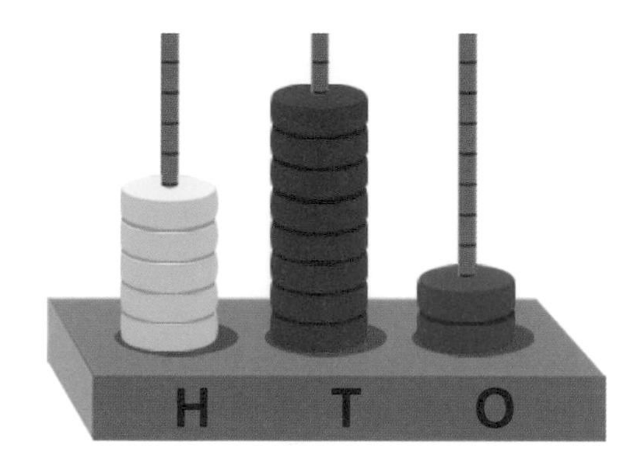

H	T	O

8 tens – ☐ tens = ☐ tens

582 – ☐ = ☐

2 Shawn makes the same number on each abacus.

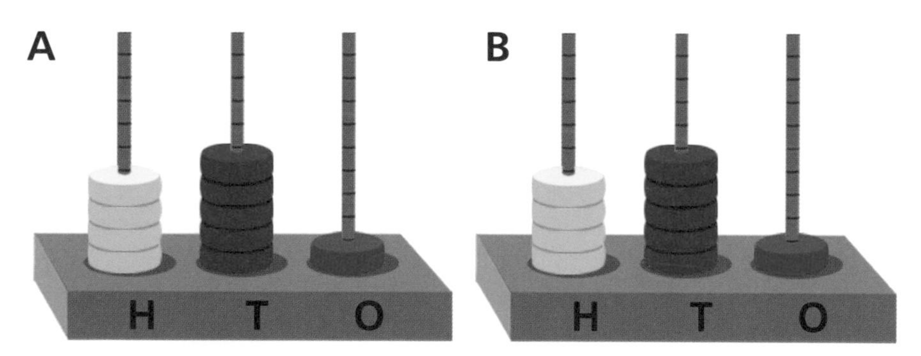

He takes 4 beads from the 10s pole of abacus **A**.

Then he places them on the 10s pole of abacus **B**.

What number does each abacus show now?

A ☐ ◯ ☐ = ☐   Abacus **A** shows ☐.

B ☐ ◯ ☐ = ☐   Abacus **B** shows ☐.

CHALLENGE

3 Match each calculation to the part-whole model that helps solve it.

Some part-whole models may be used to solve more than one calculation.

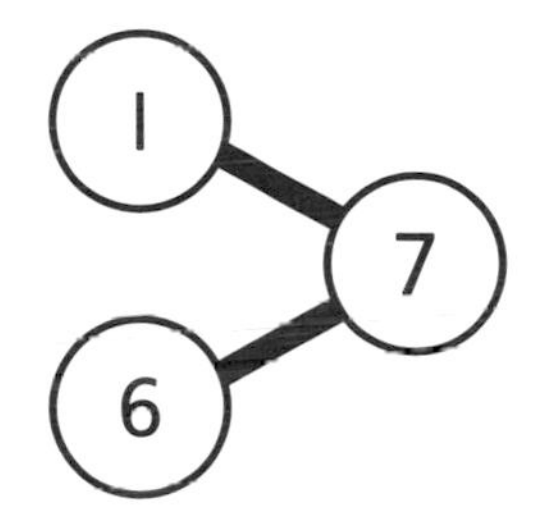

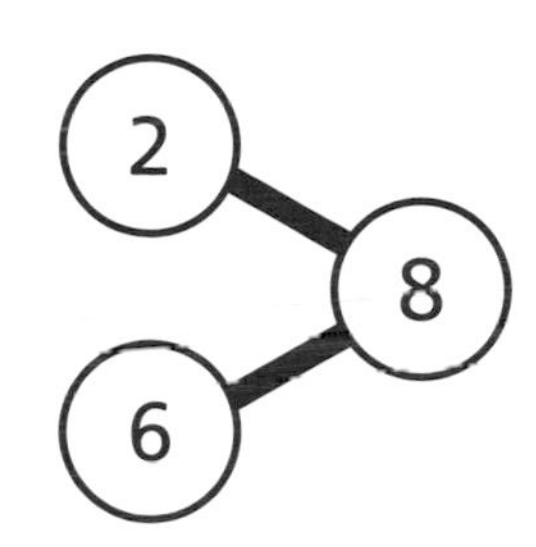

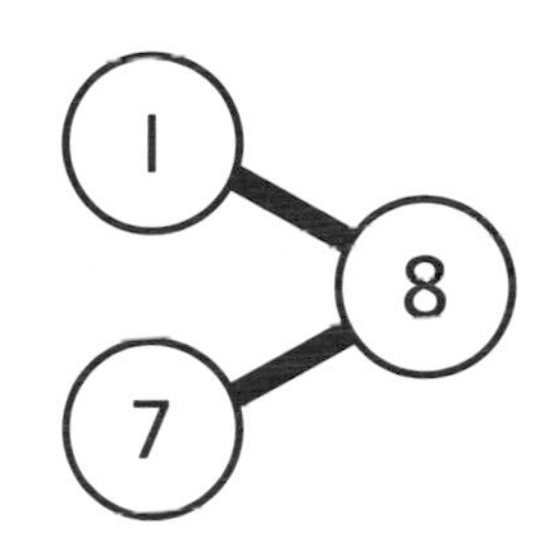

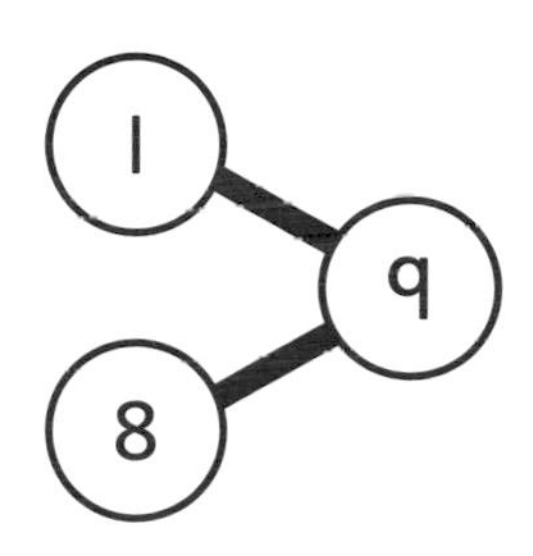

414 + 70   124 + 60   280 – 10

575 – 60   382 + 10   990 – 80

Some of these numbers have a 0 in the ones column. Does that affect the method?

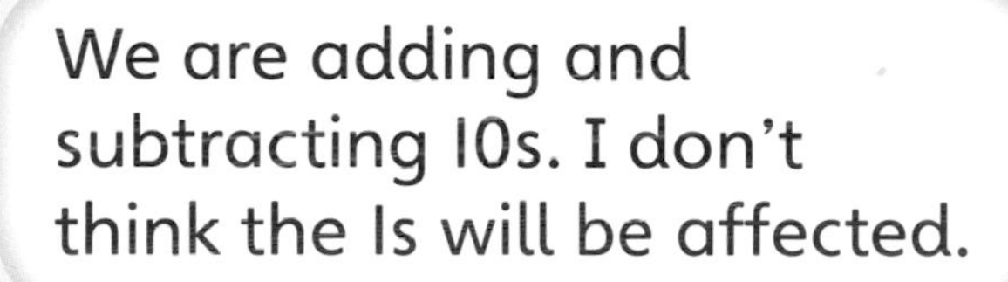

We are adding and subtracting 10s. I don't think the 1s will be affected.

→ Practice book 3A p54

Adding a 3-digit number and 10s

Discover

1 **a)** The birch tree is 10 years older than the beech tree.

The horse chestnut is 20 years older than the beech tree.

Use addition to work out their ages.

b) Represent the addition for the horse chestnut on a number line.

Share

a) 184 + 10 can be solved like last lesson but 184 + 20 is different.

8 tens + 1 ten = 9 tens

184 + 10 = 194

The birch tree is 194 years old.

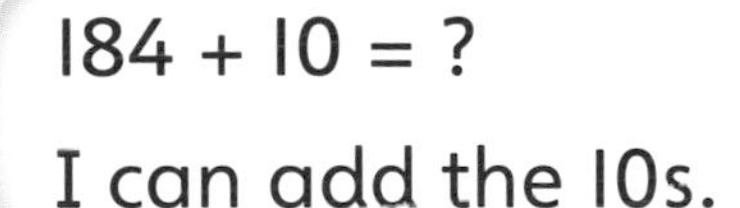
184 + 10 = ?
I can add the 10s.

H	T	O

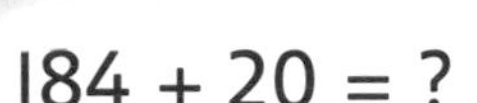
184 + 20 = ?

I will add the 10s.

8 tens + 2 tens = 10 tens

There are 10 tens. I don't think 1,104 is correct.

H	T	O

Exchange 10 tens
for 1 hundred.

184 + 20 = 204

The horse chestnut tree is 204 years old.

b)

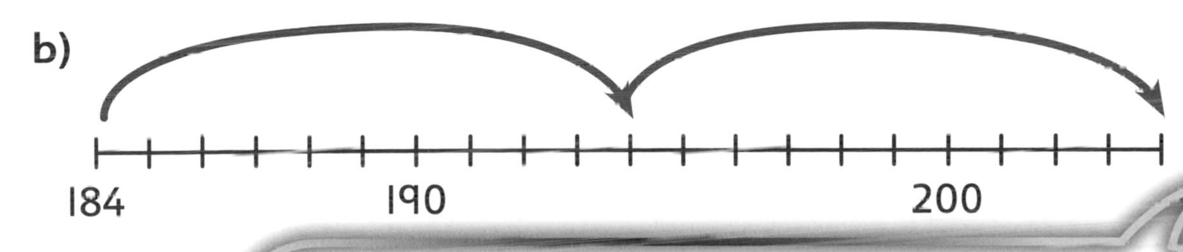
184
190
200

I wonder if I could add 16 and then add 4? But that wouldn't be as clear as adding 2 tens.

Think together

1 The oak tree is 50 years older than the beech tree.

How old is the oak tree?

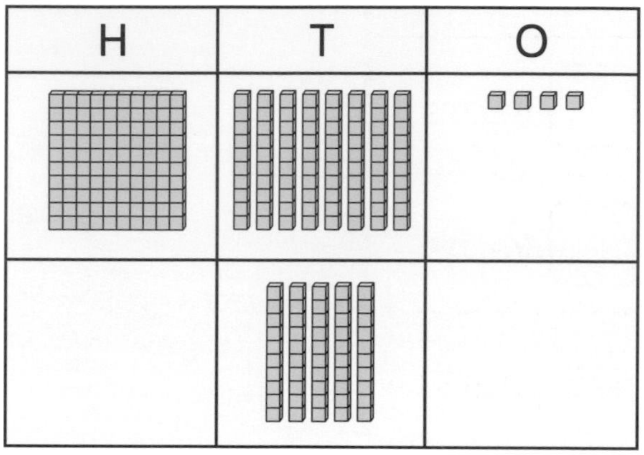

H	T	O

184 + 50 = [ ]

The oak tree is [ ] years old.

2 A giant redwood tree is 260 years old.

How old will it be in 90 years?

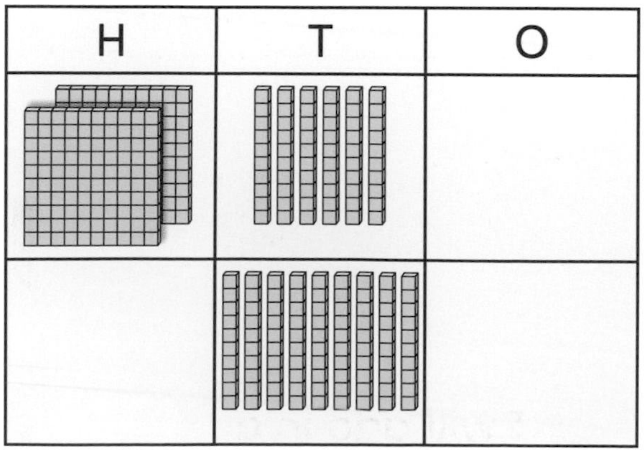

H	T	O

260 + 90 = [ ]

The giant redwood will be [ ] years old.

3 A cypress tree is 385 years old.

Complete the information in the table.

Time	Calculation	Age of tree
Present day	385 + 0	☐ years
30 years from now	385 + 30	☐ years
60 years from now	385 + ☐	☐ years
☐ years from now	385 + ☐	475 years

It looks like one of these is a missing number problem.

CHALLENGE

4 For which of these calculations do you need to exchange 10 tens for 1 hundred?

+30

181 ?

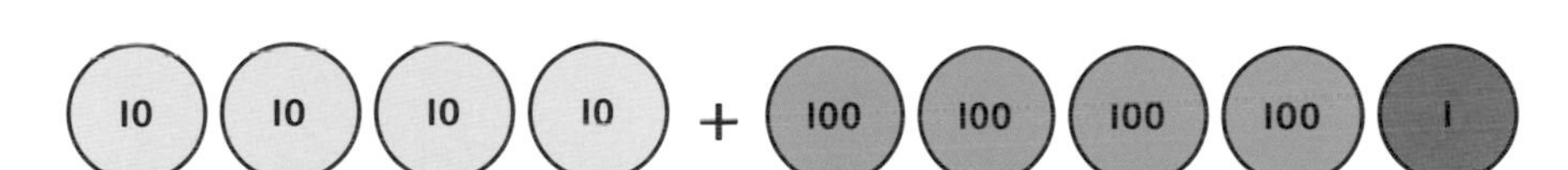

I know what to do for 365 + 60, but what about 60 + 365?

I will add in a different order.

→ Practice book 3A p57

Subtracting 10s from a 3-digit number

Discover

1 **a)** Jen has 210 m of dinosaur fabric to sell.

How much is left after she sells 20 m?

b) Jen sells some more dinosaur fabric. Now she has 140 m left.

How much did she sell?

Share

a) 210 – 20 = ?

H	T	O

I will try subtracting the tens.

20 – 10 = 10

So is the answer 210 m left?

That can't be correct; that's what I started with.

You have to exchange 1 hundred for 10 tens.

11 tens – 2 tens = 9 tens

210 – 20 = ☐

Jen has 190 m of dinosaur fabric left.

H	T	O

b) 190 – ☐ = 140

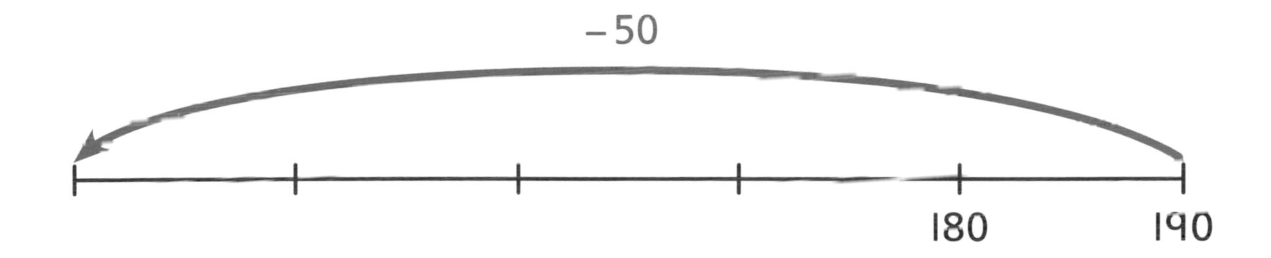

Jen has sold 50 m more dinosaur fabric.

Think together

1 Jen has 335 m of space fabric and sells 50 m.

How much is left?

H	T	O

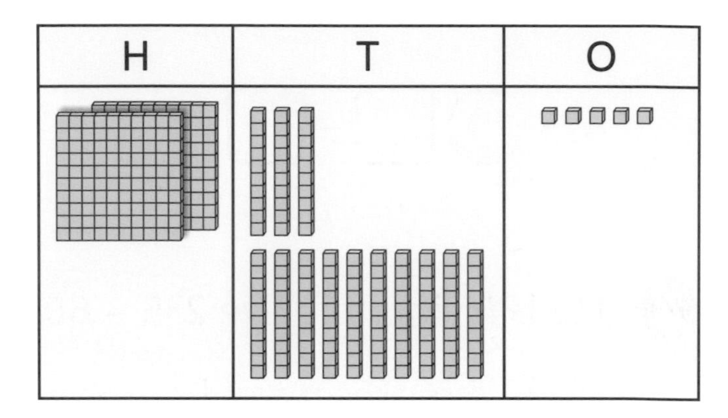

☐ tens – ☐ tens = ☐ tens

335 – 50 = ☐

There is ☐ m of space fabric left.

2 Toshi has 80 m of bee fabric to sell. Jen has 213 m of bee fabric to sell.

How much more bee fabric does Jen have than Toshi?

H	T	O

I think this is a find the difference. I can use subtraction.

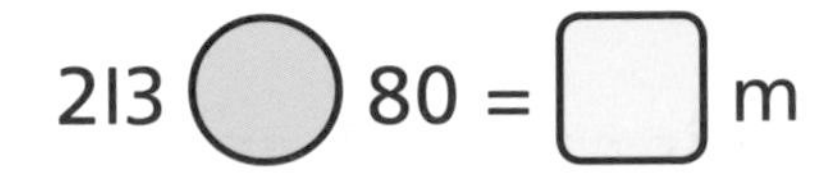

m

Jen has ☐ m of bee fabric more than Toshi.

3 What calculation does the number line show?

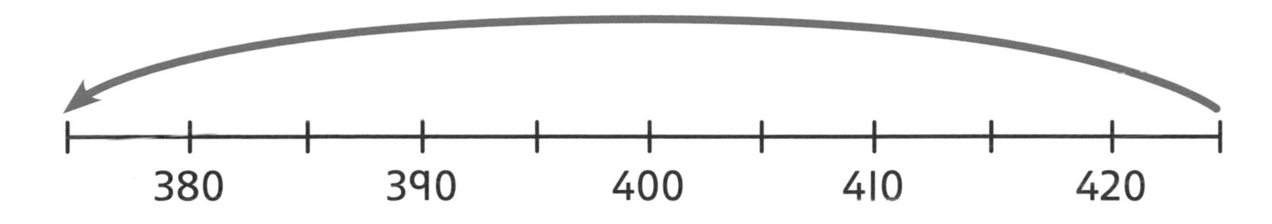

425 ◯ ☐ = ☐

4 Flo is trying to solve 235 – 60.

Flo has represented her exchange using a part-whole model.

Explain the calculation and the method used here.

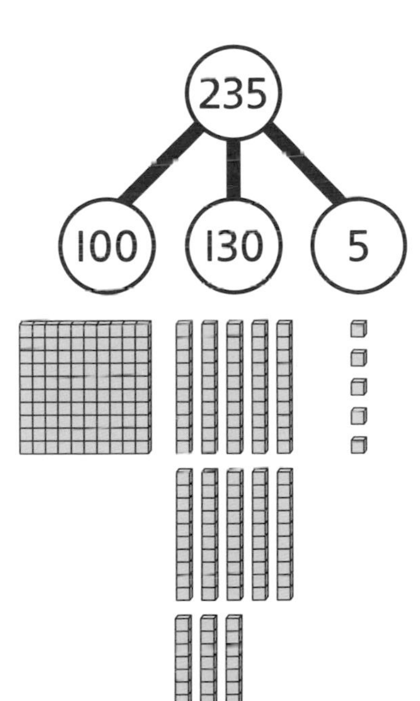

My part-whole model is a different way of showing 235.

Think of your own word problem to go with this calculation.

→ Practice book 3A p60

Adding and subtracting a 3-digit and a 2-digit number

Discover

1 **a)** Holly drives from Leicester to Bath. She then drives to Weston-super-Mare.

How many miles is Holly's total journey?

b) She has driven 11 miles from Leicester.

How far left to Bath?

Share

a) Now we are adding 10s **and** 1s.

First we add the 1s.

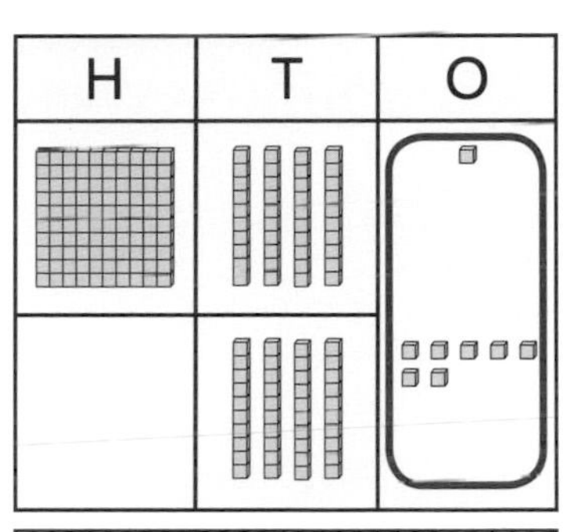
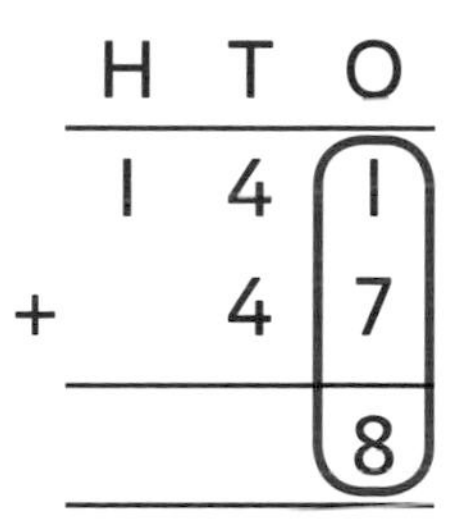

Then we add the 10s.

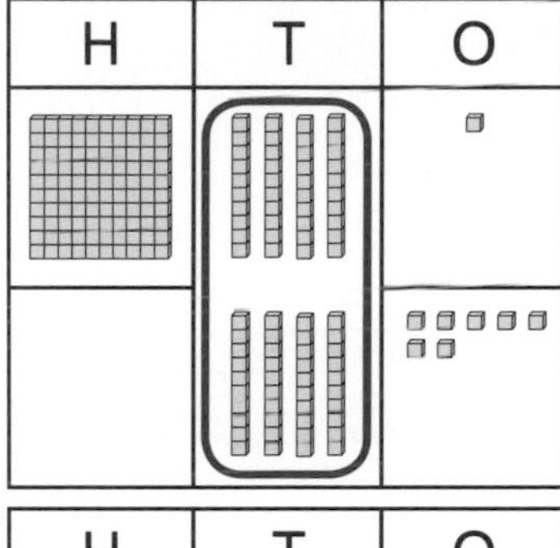
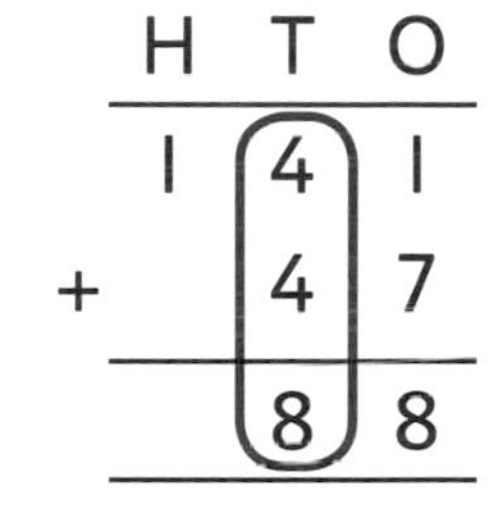

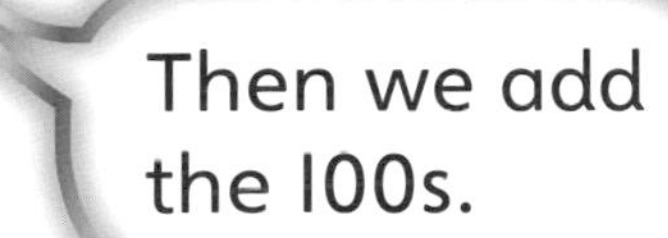

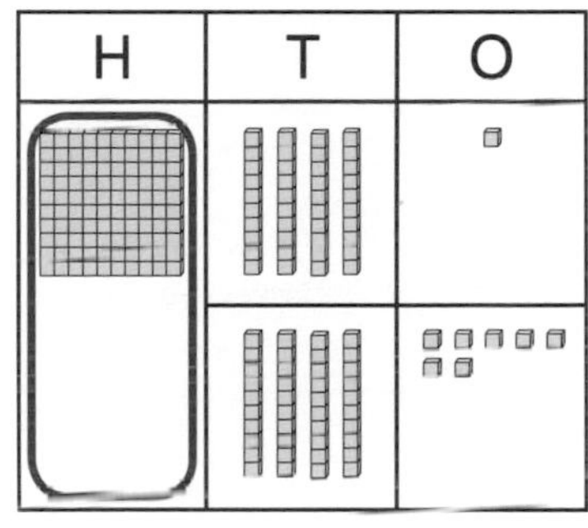

	H	T	O
	1	4	1
+		4	7
	1	8	8

141 + 47 = 188

Holly's journey is 188 miles in total.

b) It is 141 miles from Leicester to Bath. Holly has travelled 11 miles.

First subtract 1 from the ones.

Then subtract 1 ten.

141 − 11 = 130

It is still 130 miles to Bath.

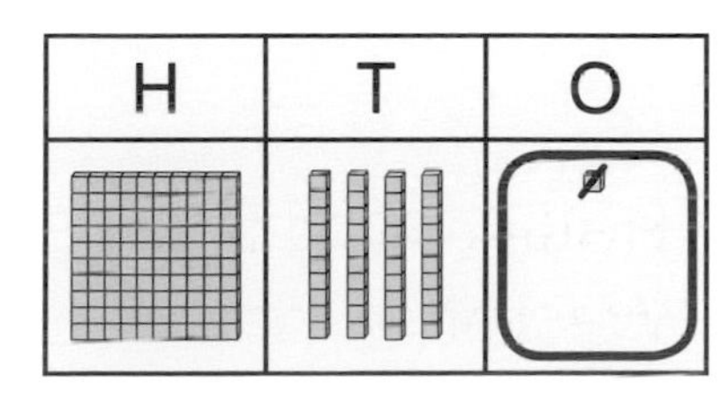
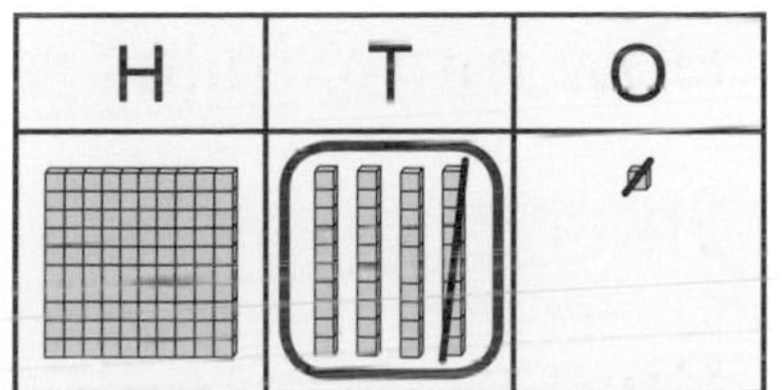

Think together

1 Jack has already driven 74 miles on the way to Cardiff from Norwich.

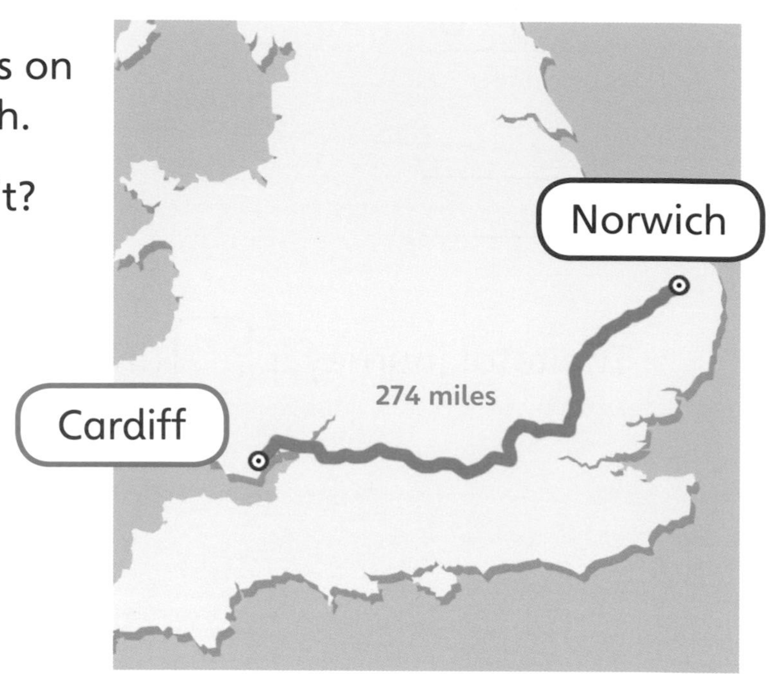

How much of the journey is left?

H T O

–

There are ☐ miles left of the journey.

2 Kay has driven 21 miles already.

How much of the journey is left?

H	T	O

H T O

– 2 1

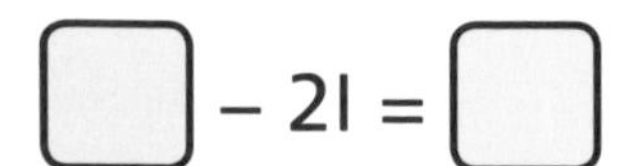

☐ – 21 = ☐

☐ miles are left.

3 How far is the total journey?

H	T	O

The total journey is ☐ miles.

4 Discuss the different methods you would use to solve these calculations.

130 + 51   609 – 7   48 + 431   234 + 52

891 – 60   938 – 26   205 + 50

I would write columns

I would solve mentally

I would check using equipment

I can do some of these mentally.

I would check using equipment or write the calculation down.

→ Practice book 3A p63

Adding a 3-digit and a 2-digit number

Discover

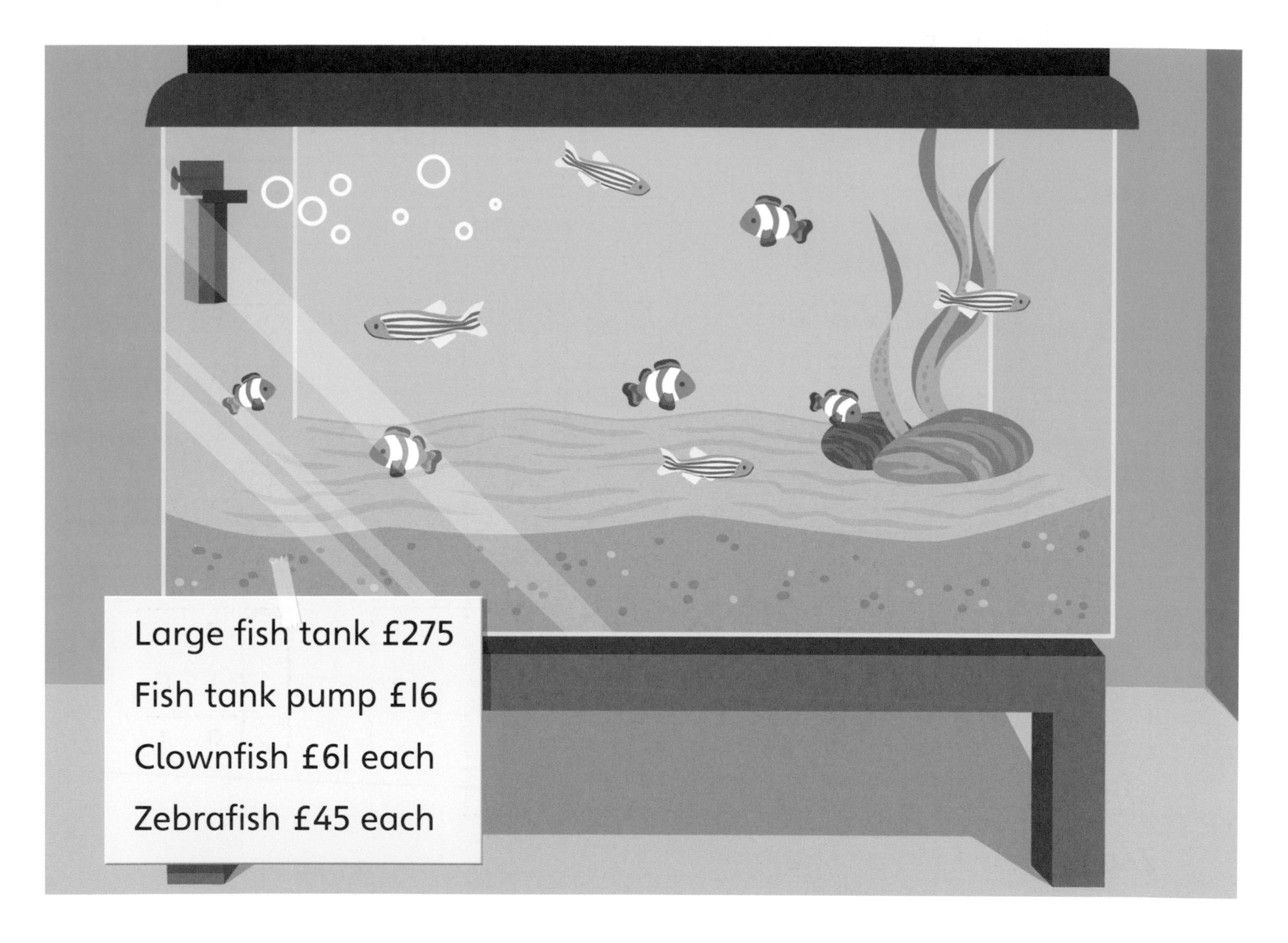

1 a) Zoe buys a large fish tank and a pump.

How much does Zoe spend altogether?

b) Aaron buys a zebrafish and a clownfish.

How much does this cost in total?

Share

a)

First add the 1s. Exchange 10 ones for 1 ten.

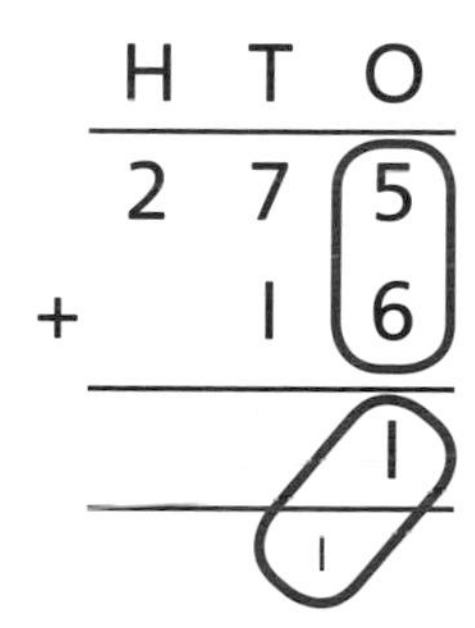

Then add the 10s. Don't forget the exchanged 10.

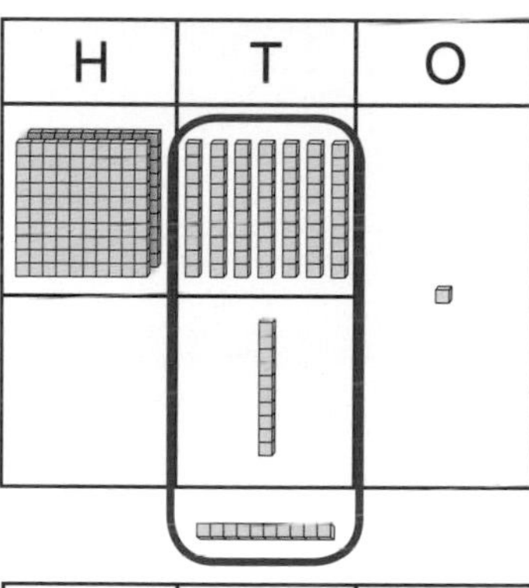

H	T	O
2	7	5
+	1	6
	9	1
	1	

And then add the 100s.

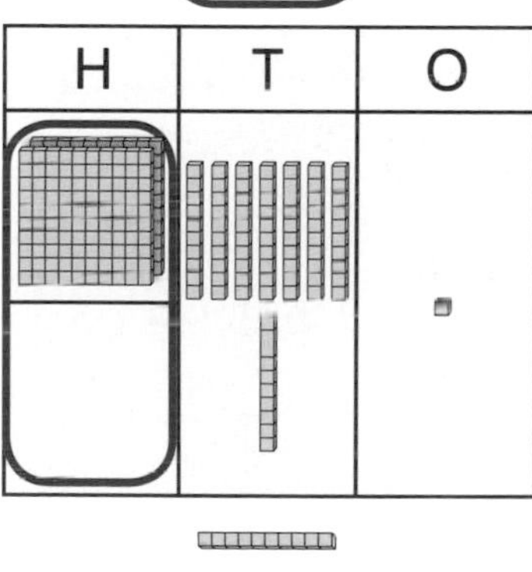

H	T	O
2	7	5
+	1	6
2	9	1
	1	

275 + 16 = 291

Zoe spends £291 altogether.

b) 45 + 61 = 106

H	T	O
	4	5
+	6	1
1	0	6

The zebrafish and the clownfish cost £106 in total.

Think together

1 Tia buys a large fish tank and a clownfish.

How much does she spend?

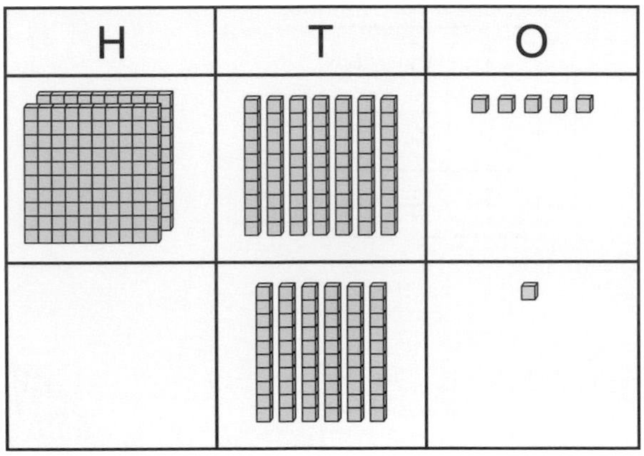

H	T	O

	H	T	O
	2	7	5
+		6	1

275 + 61 = ☐

Tia spends £☐ in total.

2 Solve these additions: 126 + 57 and 156 + 27.
What do you notice? Can you explain?

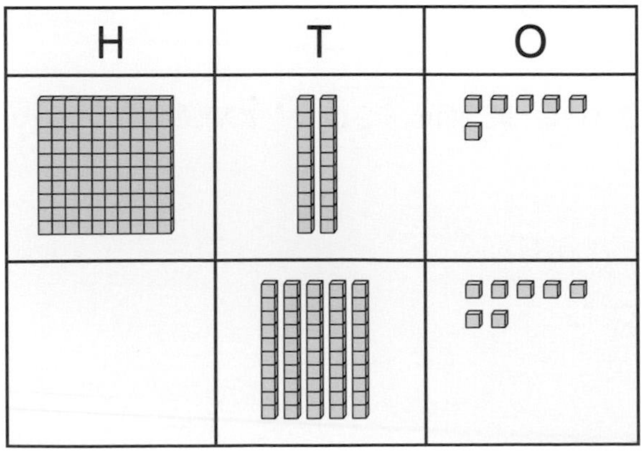

H	T	O

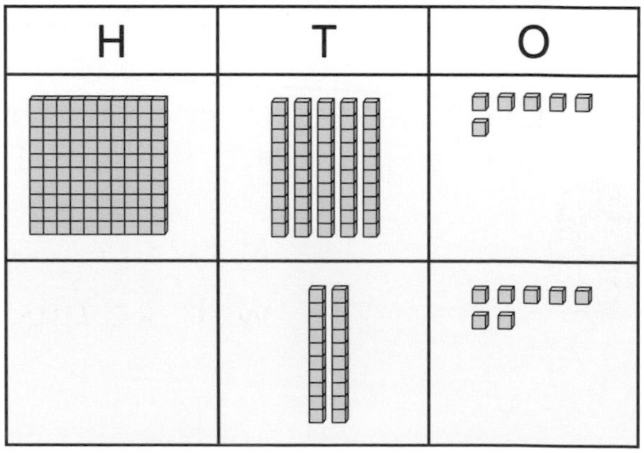

H	T	O

	H	T	O
	1	2	6
+		5	7

	H	T	O
	1	5	6
+		2	7

3 Mark and Poppy wanted to write their additions in columns.

What mistakes did they make?

Mark's addition

	H	T	O
	1	5	4
+		7	2
	1	2	6

Poppy's addition

	H	T	O
	1	6	4
+		2	7
	1	8	11

CHALLENGE

4 How many different additions can you make using these cards?

8

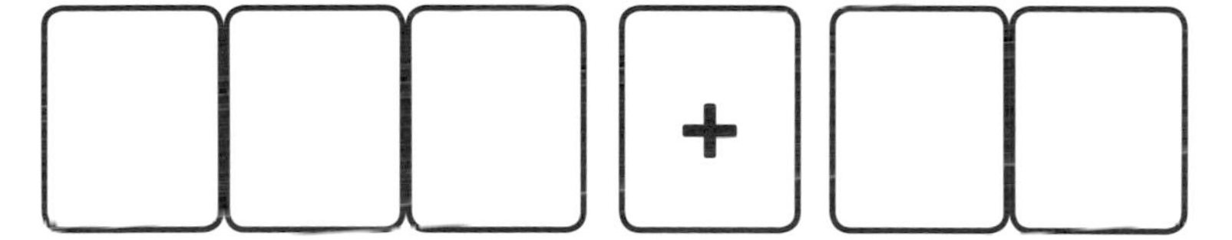

Do any of your calculations add to the same total? Explain why.

Now I can solve any addition with 3 digits and 2 digits.

I will make up some additions where you exchange for both 10s and 100s.

→ Practice book 3A p66

Subtracting a 2-digit number from a 3-digit number

Discover

1 **a)** How many new trees survived?

Luis worked it out this way. What mistake did he make?

	H	T	O
	1	7	5
−		3	8
	1	4	3

7 tens − 3 tens = 4 tens

8 ones − 5 ones = 3 ones.
So 175 − 38 = 143.

Luis

b) What is the correct answer?

Share

a)

Luis has subtracted the Is in the wrong order.

Luis should subtract 8 ones. He needs to exchange a ten.

H	T	O

b) 175 – 38 = ☐

You should first subtract 8 ones.
Then subtract 3 tens.

H	T	O

H	T	O

I will write it as columns. I wonder how to show the exchange of I ten.

	H	T	O
	1	${}^{6}\not{7}$	${}^{1}5$
–		3	8
	1	3	7

175 – 38 = 137

Think together

1 Next autumn, they plant 246 new trees. 63 of them are blown down.

How many are left?

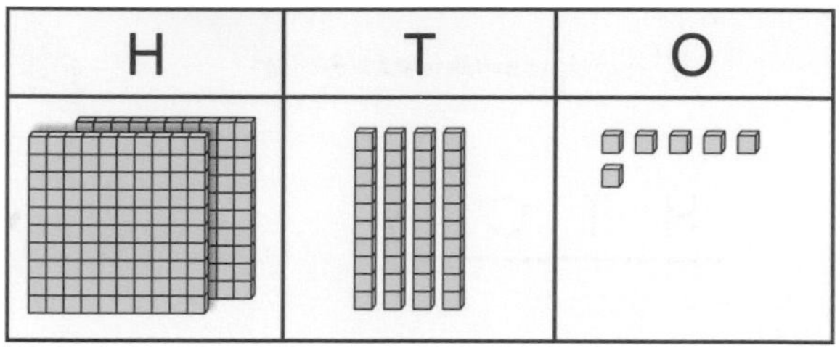

H	T	O

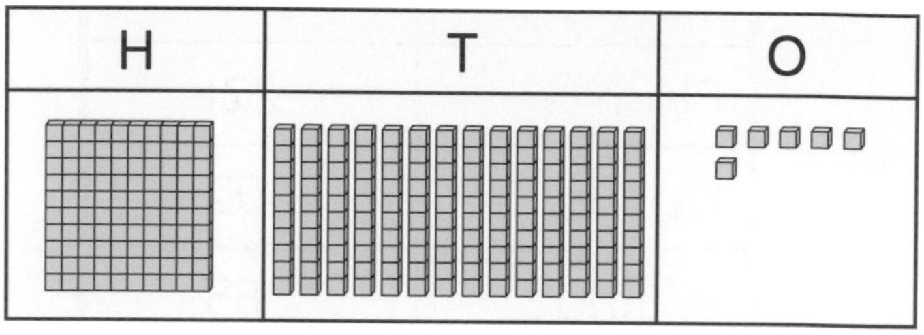

H	T	O

246 − 63 = ☐

☐ trees are left.

	H	T	O
	2	4	6
−		6	3

2 They planted 55 oak trees and 191 birch trees.

How many more birch did they plant?

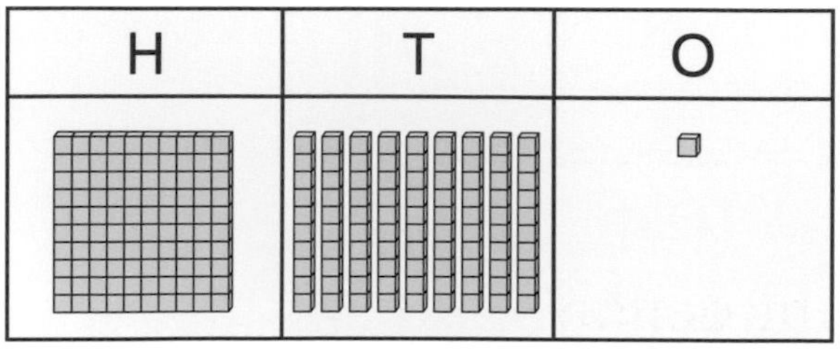

H	T	O

	H	T	O
−		5	5

They planted ☐ more birch trees.

3 How many more plum than apricot trees did they plant in the orchard?

Trees	Number planted
Apricot	43
Plum	221
Apple	302
Walnut	65

I am not sure whether to exchange a 10 or a 100 first.

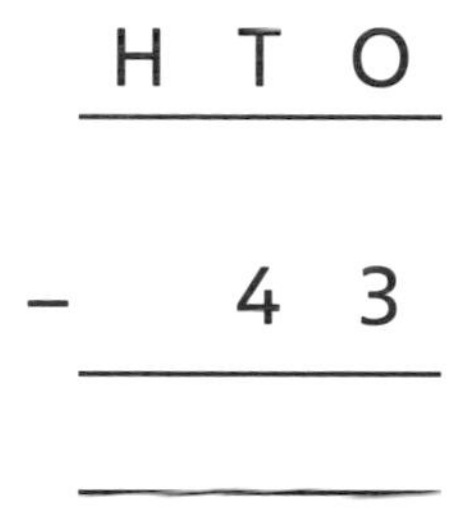

4 Flo is trying to work out 302 – 65 to find how many more apple than walnut trees have been planted.

I need to exchange a ten, but 302 has 0 tens.

H	T	O
(3 hundreds)		(2 ones)

What should Flo do to solve the subtraction?

You can check by doing an addition.

→ Practice book 3A p69

End of unit check

1 Amy scored 500 points. Ciara scored 200 points. How many more did Amy score?

A 700   **C** 500

B 300   **D** 200

Yellow arrows: Amy

Blue arrows: Ciara

2 Which calculation needs you to exchange 10 ones for 1 ten?

A 324 + 7   **B** 327 – 1   **C** 1 + 327   **D** 321 – 7

3 What is 30 less than the number shown?

2 2 5

A 255   **B** 230   **C** 195   **D** 175

4 Which calculation has an answer with a 9 in the tens column?

A 234 – 33 **B** 234 – 61 **C** 234 + 64 **D** 234 + 55

5 Which calculation does this represent?

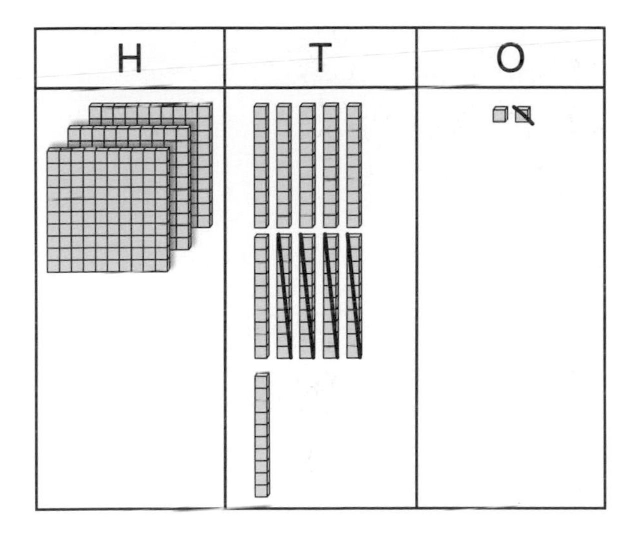

A 312 – 41 **B** 312 + 371 **C** 412 – 41 **D** 412 + 41

6 Work out the missing digits in this calculation.

2		9	=	9	5	+		5	

→ Practice book 3A p72

Unit 3
Addition and subtraction 2

In this unit we will …

- Add and subtract 3-digit numbers
- Decide if we need to exchange
- Exchange across more than one column
- Learn how to check our answers in different ways
- Use bar models to solve 1- and 2-step problems

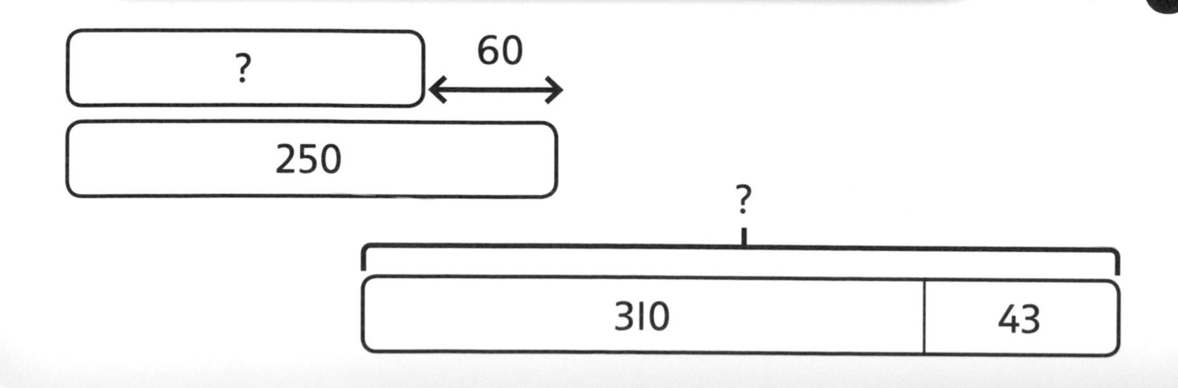

Do you remember how to find the missing information on these bar models?
?
60
250
?
310
43

We will need some maths words. Which word means to find a rough answer?

exchange **column method**

mental method

estimate **approximate**

digits **multiple**

We need to remember about parts and wholes. Use this model to find a family of 8 facts.

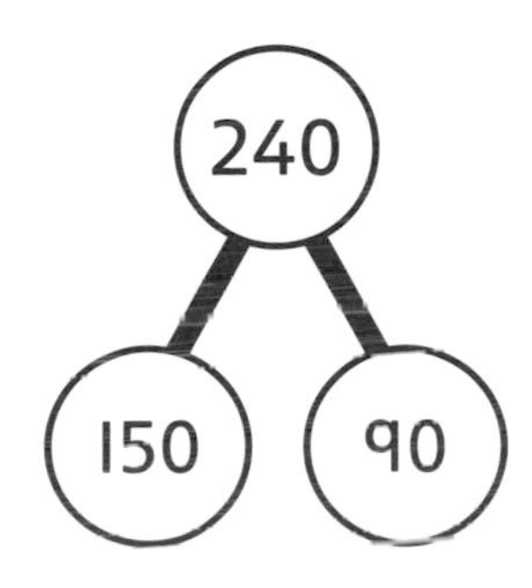

Addition and subtraction patterns

Discover

1 a) Lee inputs 154 into each function machine.

What will the outputs be?

b) Jamie inputs a number into the + 200 machine. The output is 797.

What number did she put in?

Share

a) The first machine adds 1s. The second adds 10s. The third adds 100s.

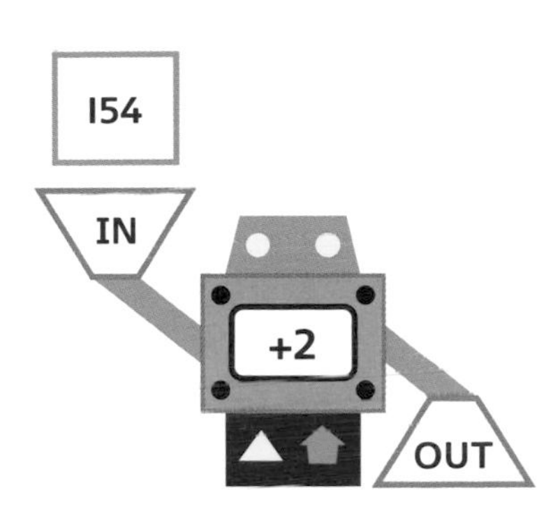

H	T	O

	H	T	O
	1	5	4
+			2
	1	5	6

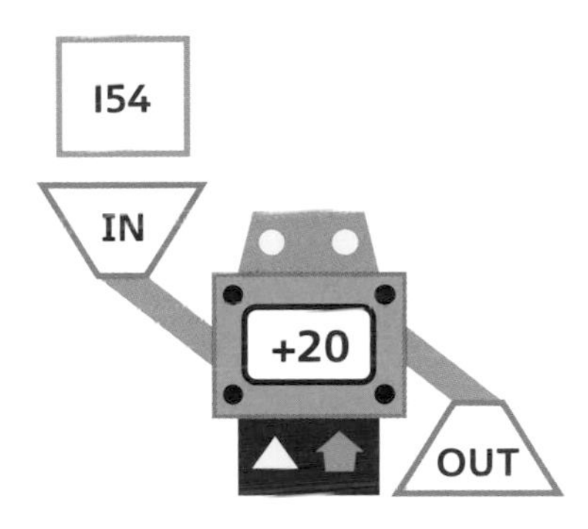

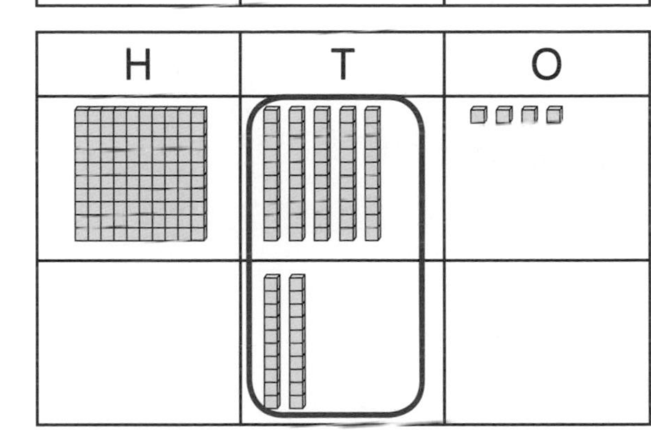

	H	T	O
	1	5	4
+		2	0
	1	7	4

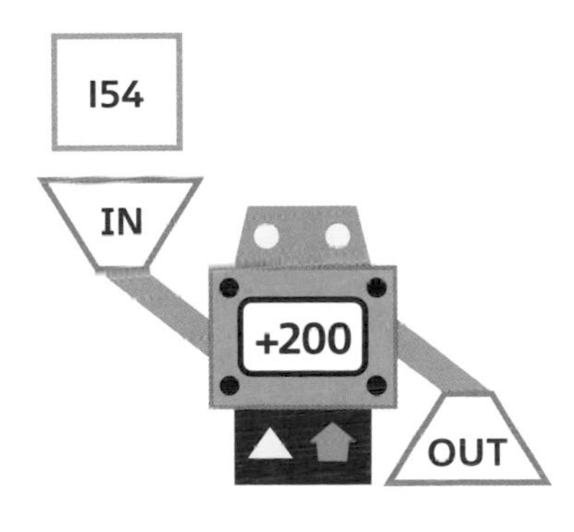

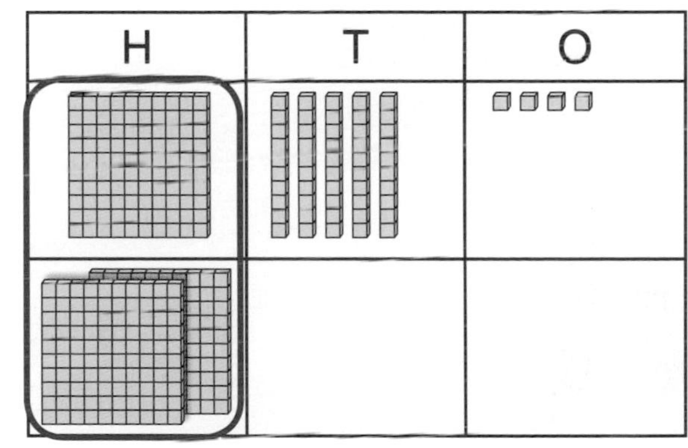

	H	T	O
	1	5	4
+	2	0	0
	3	5	4

b) ☐ + 200 = 797

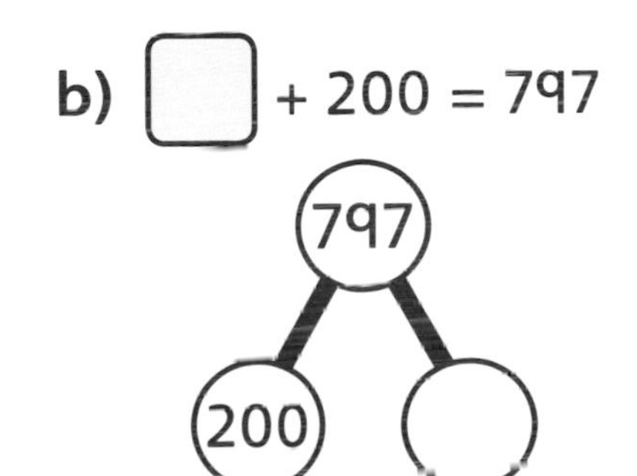

This is a missing number problem.
I will use a part-whole model to help.

797 – 200 = 597

Jamie put in number 597.

Think together

1 Find the outputs for these machines.

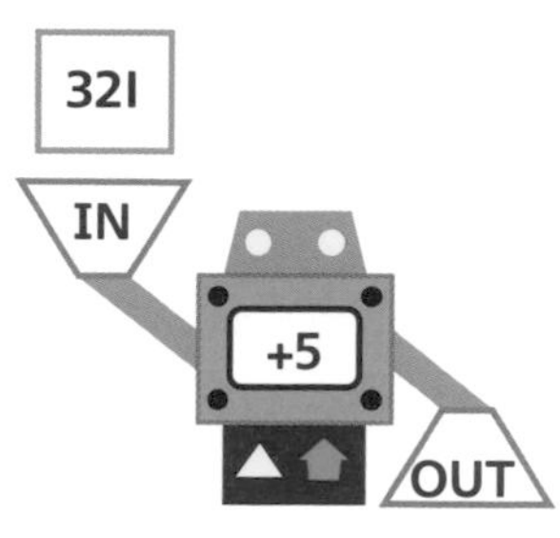

$321 + 5 = \square$

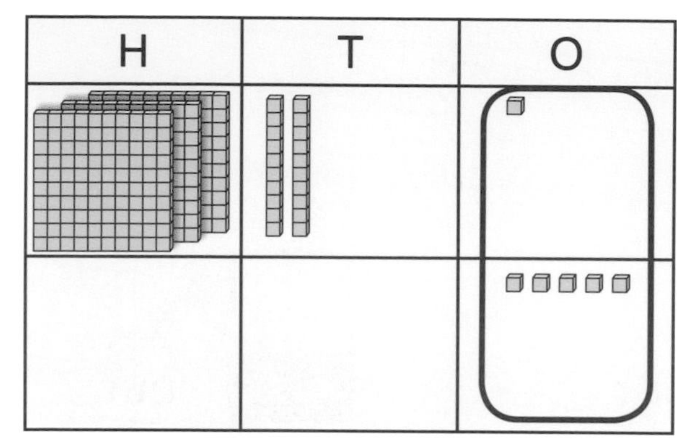

	H	T	O
	3	2	1
+			5

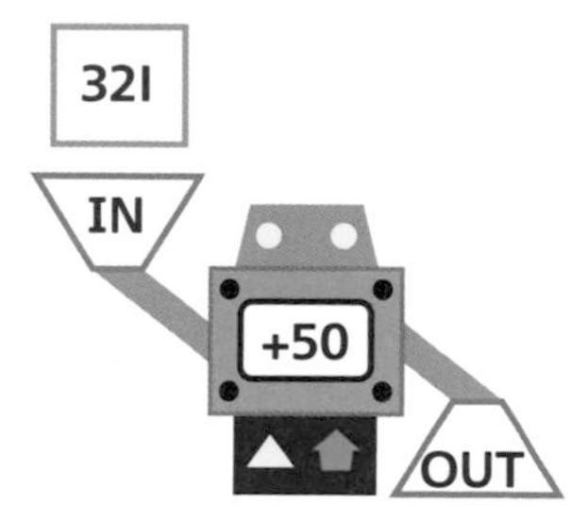

$321 + 50 = \square$

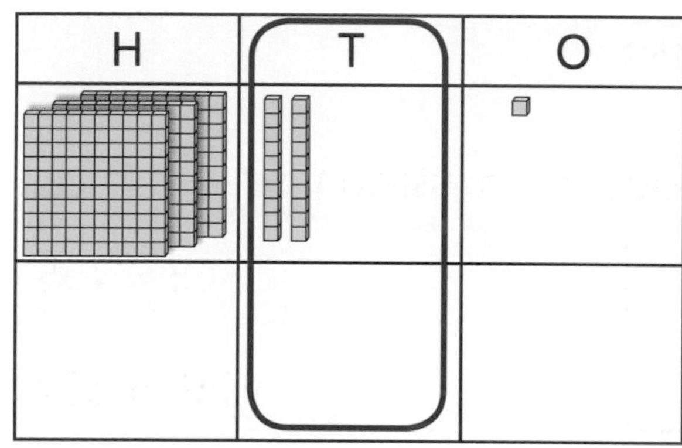

	H	T	O
	3	2	1
+			

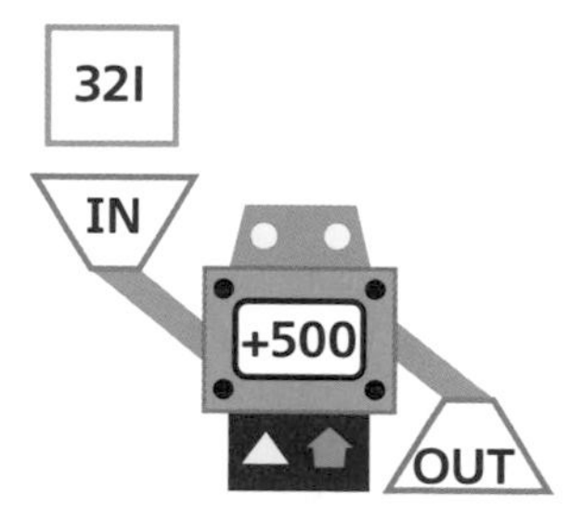

$321 + 500 = \square$

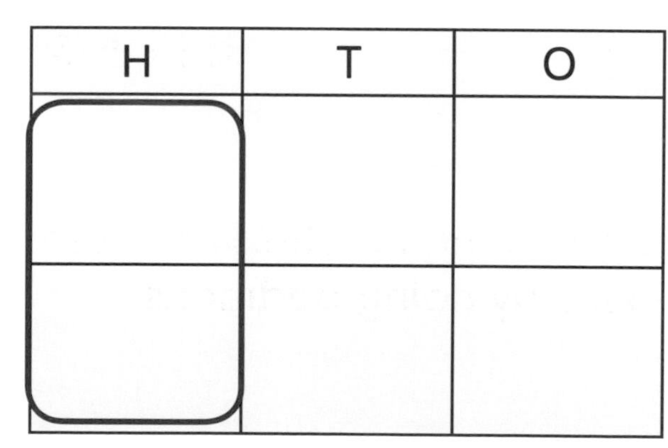

	H	T	O
	3	2	1
+			

2 546 is input into each machine. Find the missing outputs.

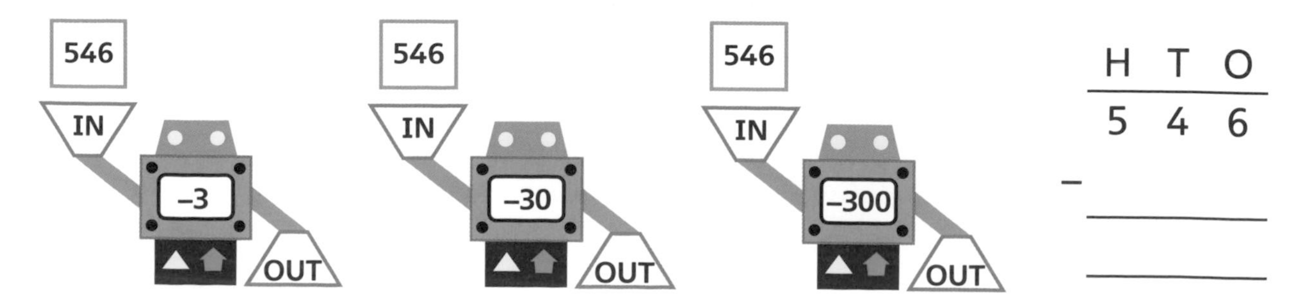

	H	T	O
	5	4	6
−			

3 **a)** The functions are missing from these machines.

Write the calculations to work out the missing functions.

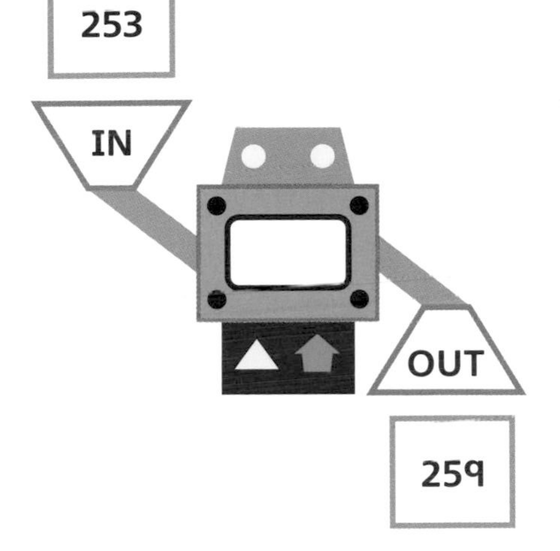

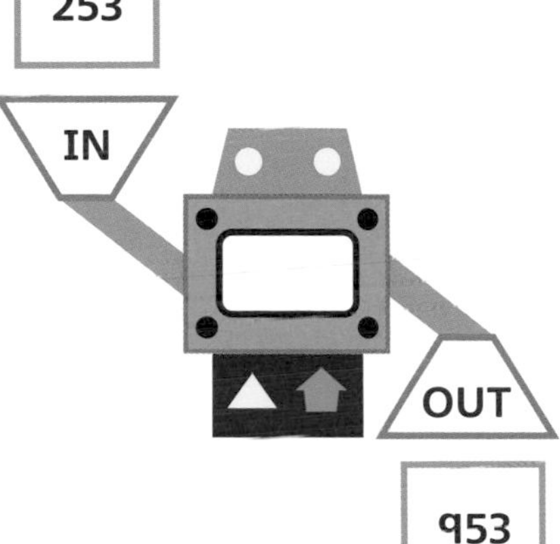

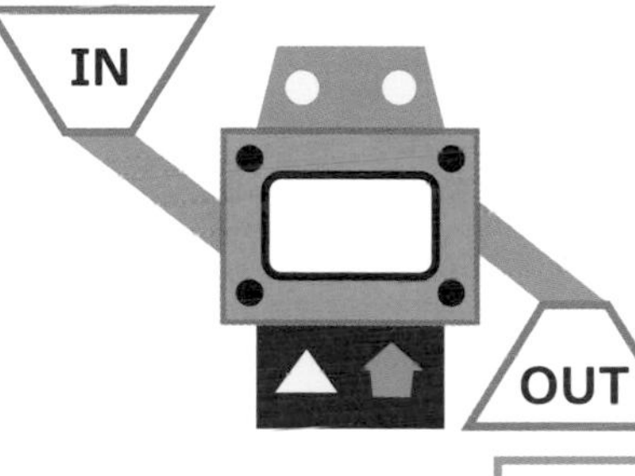

253

203

b) Now work out the missing parts of these calculations.

113 = 111 + ☐

131 = 111 + ☐

311 = 111 + ☐

555 = 755 ◯ ☐

555 = 557 ◯ ☐

555 = 575 ◯ ☐

I will guess the function and test my ideas by doing additions.

Let's try to work it out by looking at which **digits** change.

→ Practice book 3A p74

Adding two 3-digit numbers 1

Discover

1 **a)** Richard uses digit cards to make the numbers 3 2 6 and 5 4 1.

He adds the numbers together.

What is his total?

b) Richard takes a digit card from one number and swaps it with a digit card from the other number.

His total is the same.

Which digits did he swap?

Share

a)

I can use a place value grid to organise my thinking.

3 2 6

	H	T	O
3	2	6
+	5	4	I
			7

	H	T	O
	3	2	6
+	5	4	I
		6	7

	H	T	O
	3	2	6
+	5	4	I
	8	6	7

Richard's total is 867.

b)

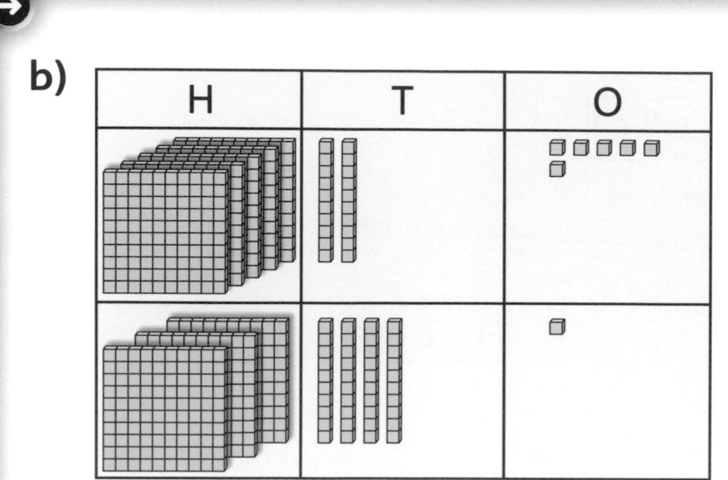

	H	T	O
	5	2	6
+	3	4	1
	8	6	7

Richard swapped the 5 and the 3.

I wonder if there are any other digits he could swap.

Think together

1 Richard makes two different numbers.

His numbers are 1 4 2 and 3 5 6.

What is his total?

	H	T	O
	1	4	2
+	3	5	6

2 Jamilla uses the digit cards to make two numbers.

Her numbers are 4 1 3 and 5 6 2 .

What is her total?

H	T	O

H T O

\+

3 Lexi gets a total of 993.

What numbers did she start with to get this total?

H T O

\+

I think there is more than one answer.

→ Practice book 3A p77

Adding two 3-digit numbers 2

Discover

1 **a)** How many birds of prey have Amal and Jen seen in total?

b) They see a further 20 male and 30 female birds of prey.

How many have they seen in total now?

Share

a) There are 126 males and 217 females.

I will use base 10 equipment but there are more than 10 ones in total.

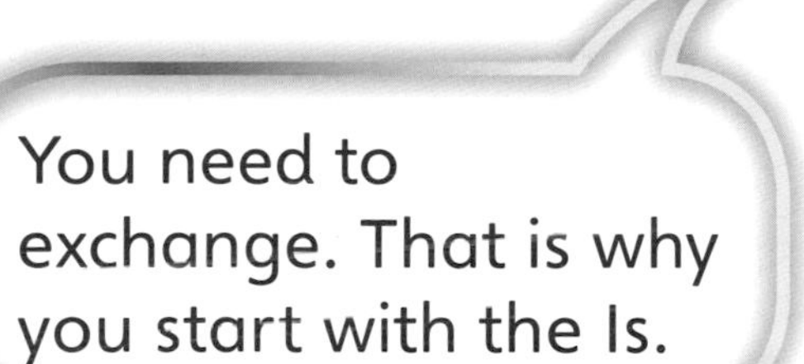

You need to exchange. That is why you start with the 1s.

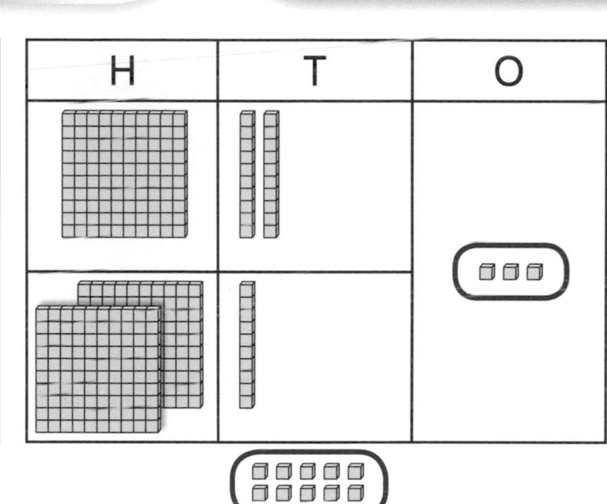

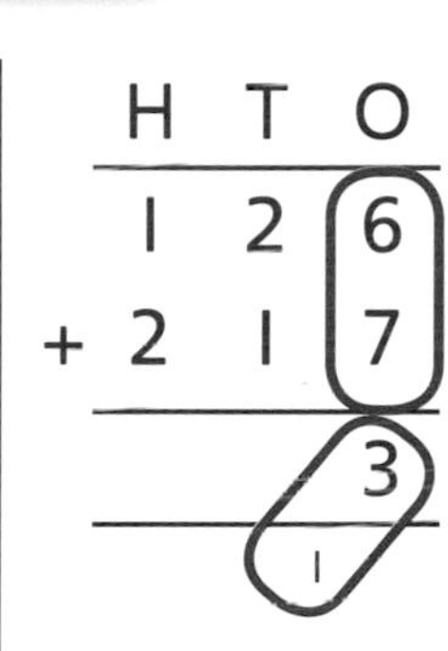

Now add the 10s. Remember to add the exchanged 10 too.

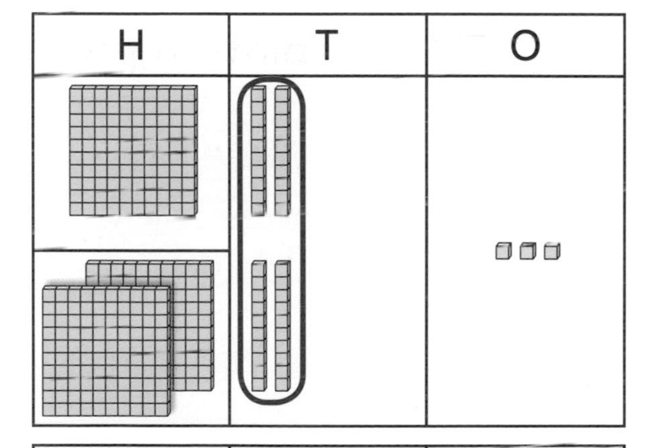

Then add the 100s.

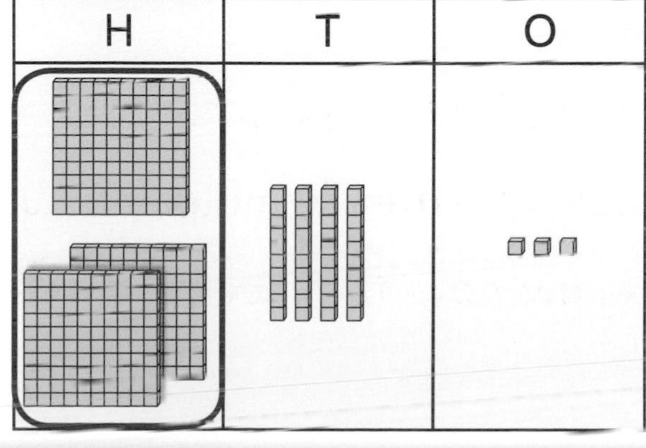

126 + 217 = 343

Amal and Jen saw 343 birds of prey in total.

b) 20 + 30 = 50. Now there are 50 more.

I did two additions. I added 20 to 343. Then I added 30 to the new total.

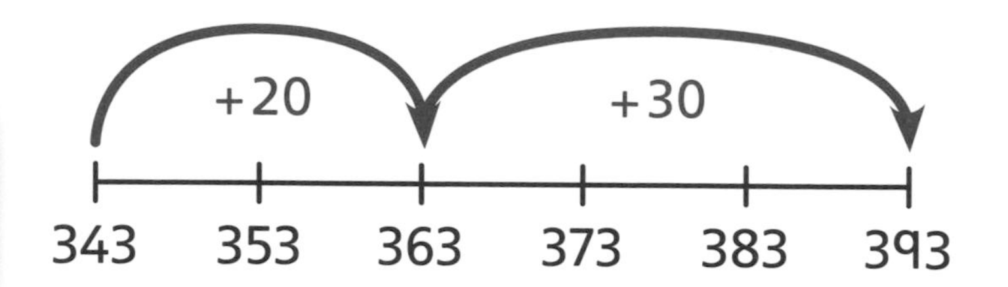
+20
+30
343
353
363
373
383
393

	H	T	O
	3	4	3
+		5	0
	3	9	3

343 + 50 = 393

They have now seen 393 birds of prey in total.

Think together

1 Amal and Jen also spot blackbirds.

They see 262 males and 251 females.

How many blackbirds are there altogether?

H	T	O

	H	T	O
	2	6	2
+	2	5	1

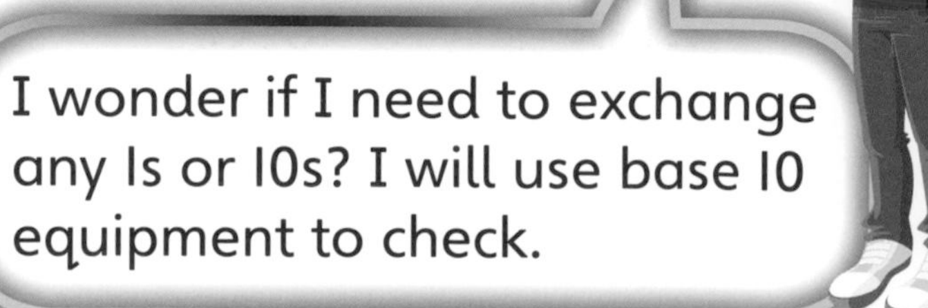
I wonder if I need to exchange any 1s or 10s? I will use base 10 equipment to check.

2 The next month, Amal and Jen spot 157 blackbirds and 166 birds of prey.

How many birds are there altogether?

H	T	O

H T O

\+

CHALLENGE

3 **a)** Max is working out 184 + 217.

I think I will only need to exchange the 1s because the 10s only add up to 9.

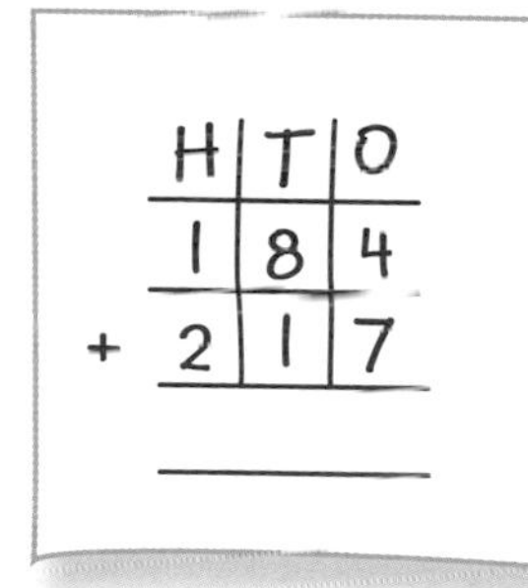

Is Max correct?

Test his idea by doing the calculation.

b) Invent three different calculations that have the same effect.

What do you notice?

→ Practice book 3A p80

Subtracting a 3-digit number from a 3-digit number 1

Discover

1 **a)** Luis spins 3, 5 and 2.

He makes the subtraction 999 – 352.

What is his score?

b) Isla spins 1, 6 and 6.

Use 1, 6 and 6 in different combinations.

What could Isla's score be?

Share

a) This is a subtraction with two 3-digit numbers.

H	T	O

	H	T	O
	9	9	9
−	3	5	2
			7

H	T	O

	H	T	O
	9	9	9
−	3	5	2
		4	7

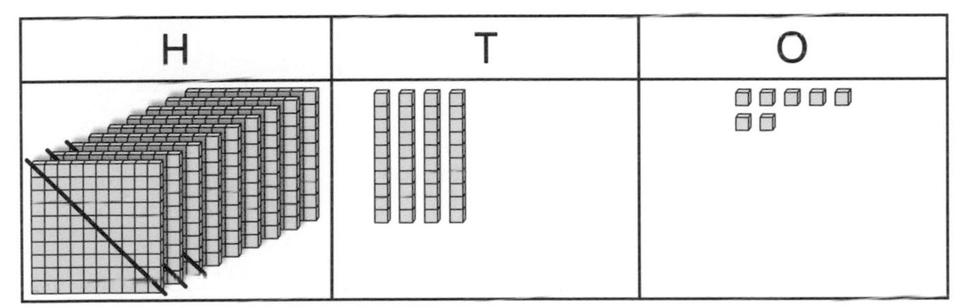

	H	T	O
	9	9	9
−	3	5	2
	6	4	7

I checked using a number line.

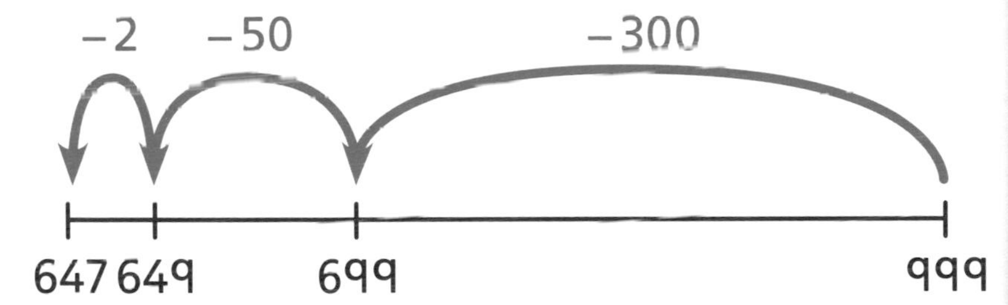

Luis scored 647.

b) Isla could score 833, 383 or 338.

	H	T	O
	9	9	9
−	1	6	6
	8	3	3

	H	T	O
	9	9	9
−	6	1	6
	3	8	3

	H	T	O
	9	9	9
−	6	6	1
	3	3	8

Think together

1 Jamilla spins 4, 3 and 5 and makes the number 435.

What is her score?

999 – ☐☐☐

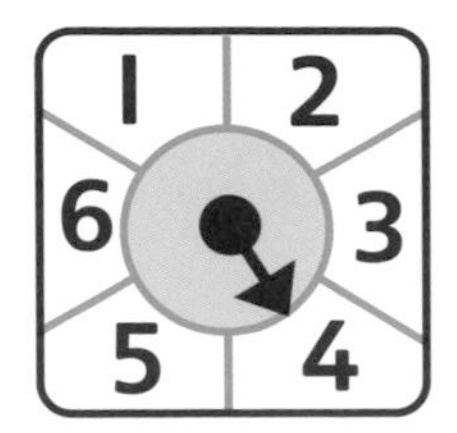

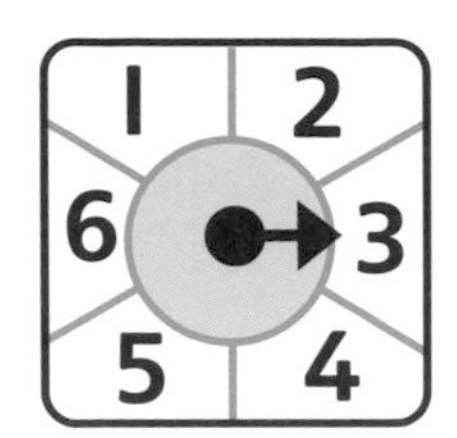

H	T	O

	H	T	O
	9	9	9
–	4	3	5

999 – 435 = ☐

Jamilla's score is ☐.

2 Ebo has the subtraction 678 – ☐☐☐

He spins 4, 4 and 6.

Find 3 different scores Ebo can make.

	H	T	O
–			

	H	T	O
–			

	H	T	O
–			

Ebo could score ☐, ☐ or ☐.

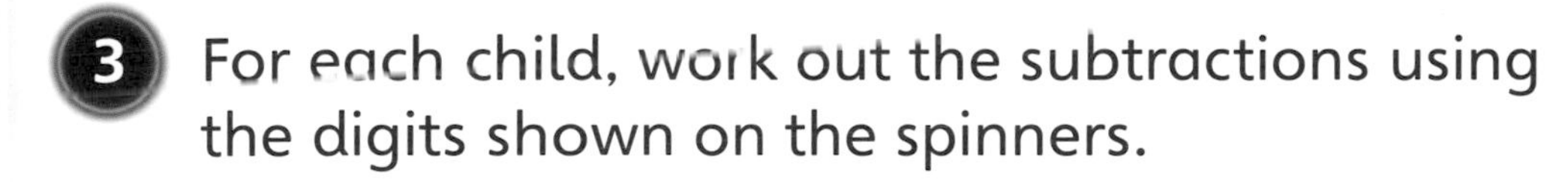

3 For each child, work out the subtractions using the digits shown on the spinners.

If suitable, use a **mental method** to get the answer.

My score is an even number.

Mo

My score is less than 100.

Reena

My score is a **multiple** of 10.

Ambika

I made an odd number greater than 500.

Andy

Is there more than one answer for each child?

I could just try some different examples.

Let's try thinking about the digits logically. Some of the subtractions I can do mentally.

→ Practice book 3A p83

Subtracting a 3-digit number from a 3-digit number ❷

Discover

There are 361 steps to the top of the tower.
145, 146, 147.
Olivia
Aki

1 **a)** How many steps does Aki have left to climb?

b) Olivia has climbed 10 more than Aki.

How many does she have left to climb?

Share

a) 361 – 147

Exchange 1 ten for 10 ones.

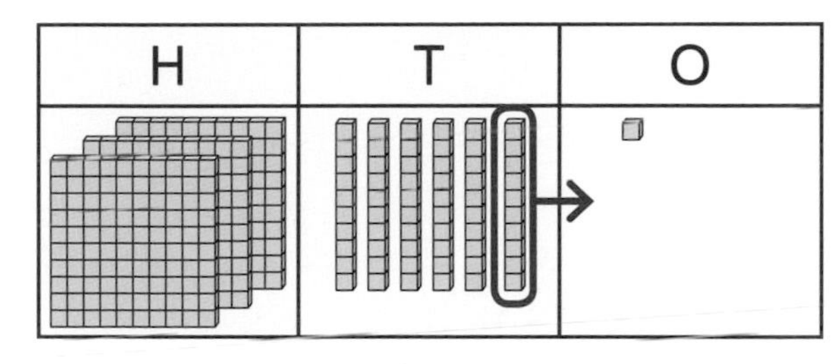

There are now 5 tens and 11 ones.

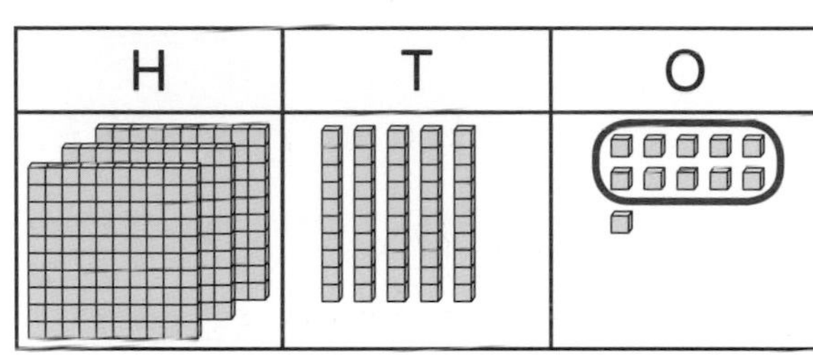

	H	T	O
	3	$^{5}\cancel{6}$	$^{1}1$
–	1	4	7

Subtract the 1s.

H	T	O
3	$^{5}\cancel{6}$	$^{1}1$
– 1	4	7
		4

Then subtract the 10s.

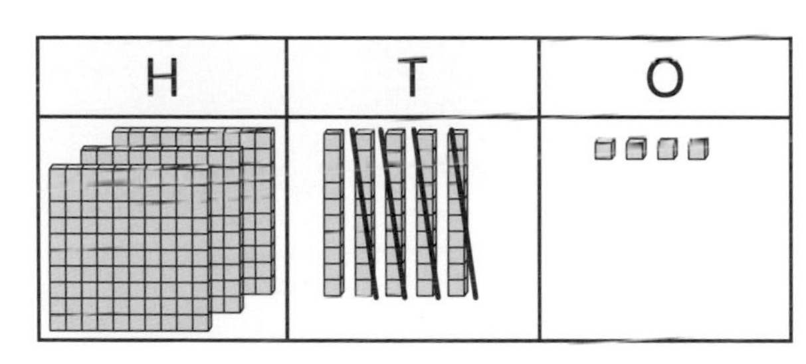

	H	T	O
	3	$^{5}\cancel{6}$	$^{1}1$
–	1	4	7
		1	4

Then subtract the 100s.

	H	T	O
	3	$^{5}\cancel{6}$	$^{1}1$
–	1	4	7
	2	1	4

361 – 147 = 214

Aki has 214 steps left to climb.

b)

Olivia has 10 fewer steps than Aki left to climb. I will work out 214 – 10 mentally.

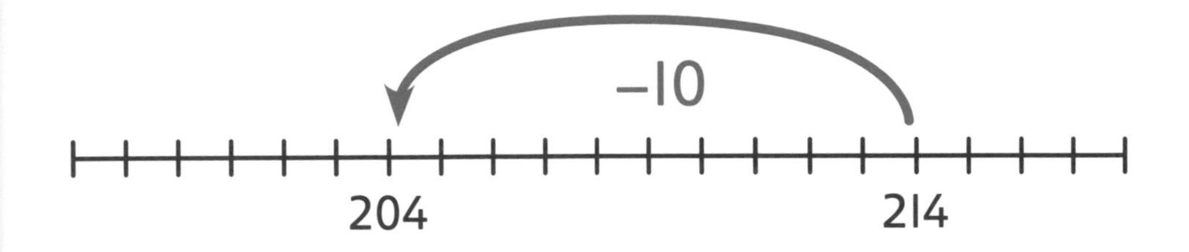

	H	T	O
	2	1	4
–		1	0
	2	0	4

214 – 10 = 204

Olivia has 204 steps left to climb.

Think together

1 There are 525 steps to the top of a skyscraper. Lexi has climbed 361.

How many steps does she still have to climb?

H	T	O

H	T	O

	H	T	O
	⁴~~5~~	¹2	5
–	3	6	1

Lexi has ☐ more steps left to climb.

2 Emma is also climbing the skyscraper.

She has 387 steps left to the top. How many has she already climbed?

	H	T	O
–			

Emma has already climbed ☐ steps.

3 These are very common mistakes.

Explain what has happened.

a) 314 – 253 = ☐

	H	T	O
	3	1	4
–	2	5	3
	1	4	1

b) ☐ = 553 – 255

	H	T	O
	5	5	13
–	2	5	5
	3	0	8

4 Discuss how to solve this subtraction.

I know I need to exchange for 10 ones, but there aren't any tens.

I need to work out how this is possible.

	H	T	O
	5	0	6
–	3	2	8

→ Practice book 3A p86

Estimating answers to additions and subtractions

Discover

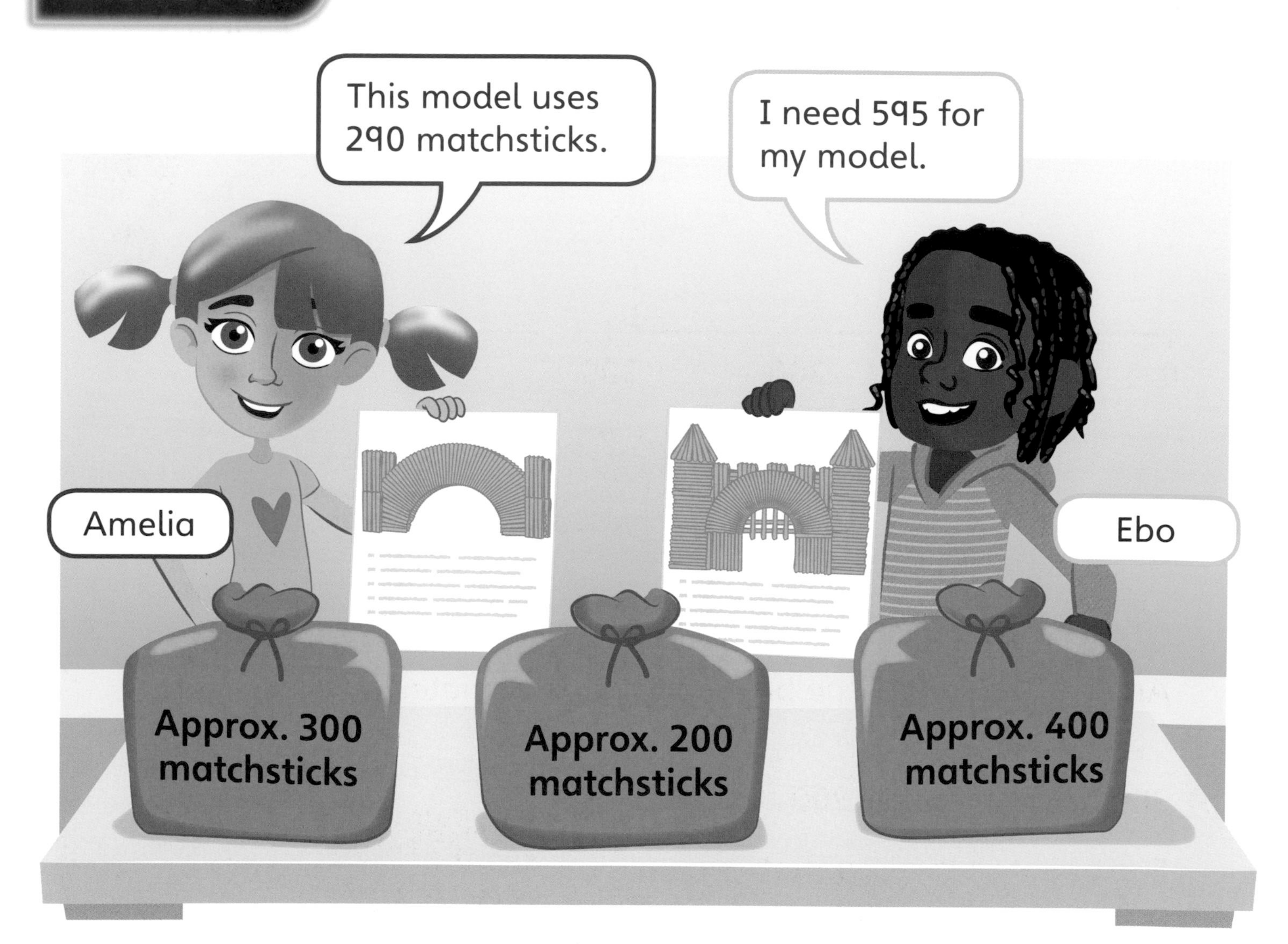

1 **a)** Which bag should Amelia pick to make her model?

b) Ebo counts all of the matchsticks from one bag.

There are exactly 211. Which bag did he count from?

Share

a) Amelia needs 290 matchsticks.

One bag has approximately 200. I think she should pick that one.

Approx. is short for **approximately**. 'Approximately 300' means a number close to 300.

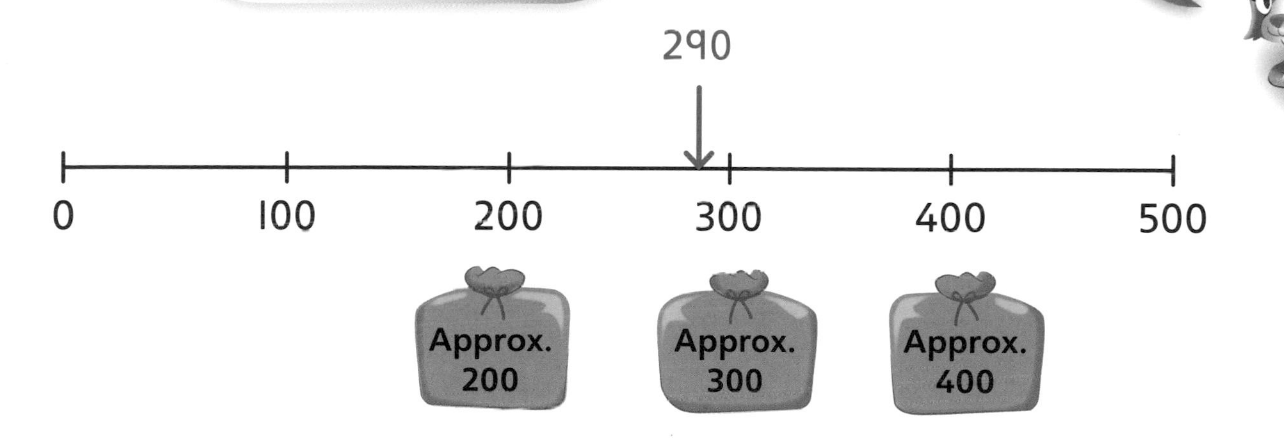

290 is approximately 300.

Amelia should pick the bag that is approximately 300.

b) 211 is approximately 200.

Ebo counted the bag that is approximately 200.

I can use a number line to see that 211 is much closer to 200 than to 300.

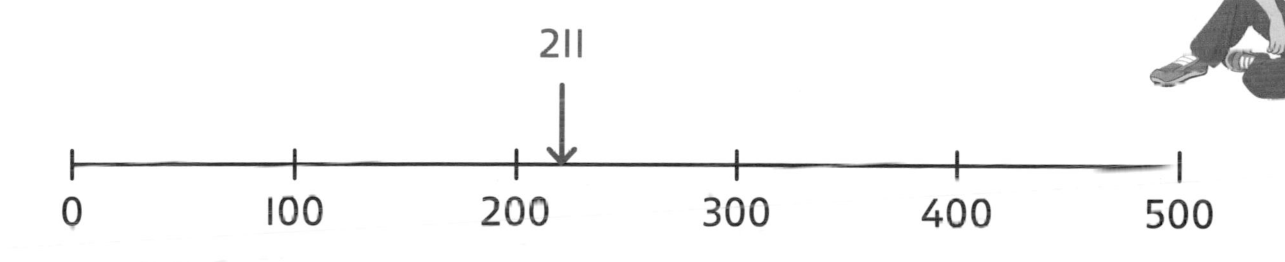

Think together

1 Estimate the numbers that are being shown by the letters on the number line.

Write each one as a sentence, such as:

'I estimate that A is about ☐.

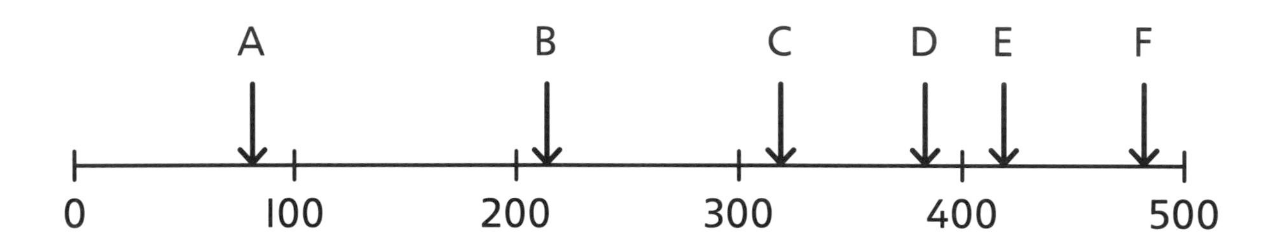

2 Show where these numbers approximately appear on the number line.

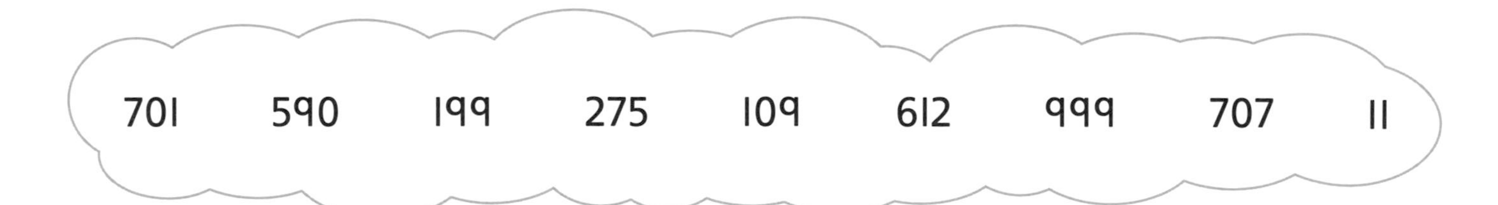

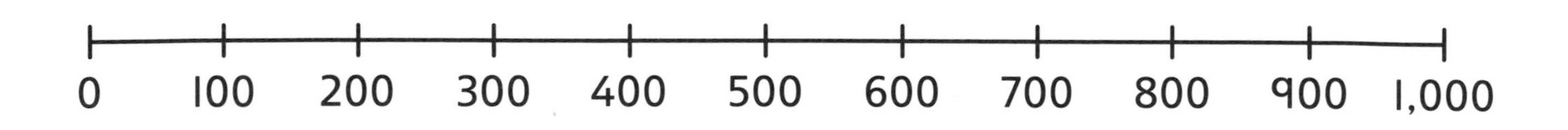

I will see which 100 each one is nearest to.

3 **a)** Find an approximate answer to 381 + 398.

I think the answer will be approximately 600, because I am adding 3 hundreds and 3 hundreds.

I think the answer should be closer to 800, because both numbers are approximately 400.

Who do you agree with?

What **approximation** could you use to estimate the answer?

□□□ + □□□ = □□□

b)

I worked out 512 – 280 = 332

Can I use approximation to check?

Alex

□00 – □00 = □00

Is Alex's answer close to your estimate?

Should she try the calculation again?

→ Practice book 3A p89

Checking strategies

Discover

1 **a)** Complete the part-whole models for Max and Ambika.

Use an addition to check their subtractions.

b) Correct any mistakes.

Share

a) 525 – 270 = 255   332 = 755 – 427

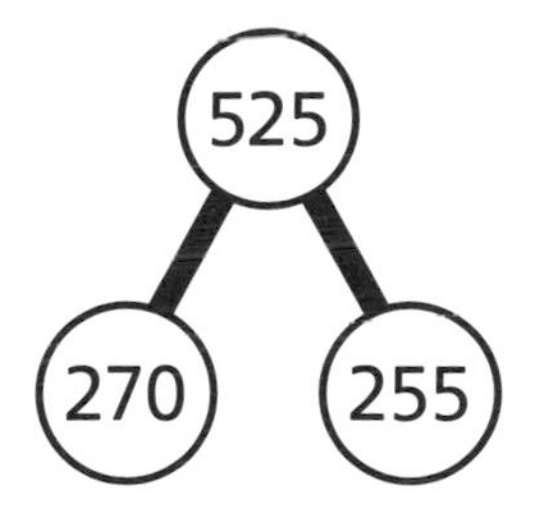

755
427 332

I will add the parts to check each calculation.

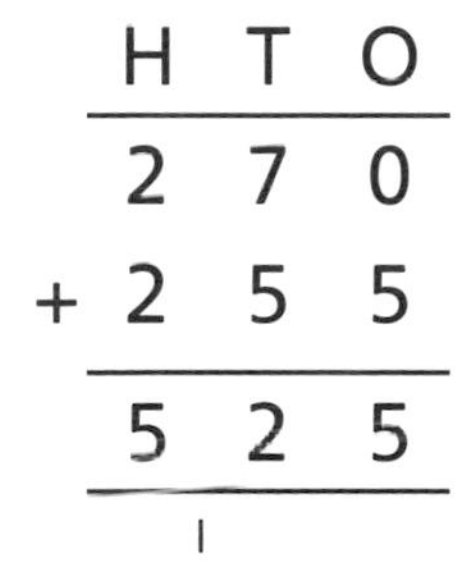

	H	T	O
	2	7	0
+	2	5	5
	5	2	5
		1	

	H	T	O
	4	2	7
+	3	3	2
	7	5	9

The addition for Max's calculation gives the same whole and parts. This shows it is correct.

Ambika's subtraction is incorrect because when you add the parts you get a different whole.

b) We need to correct Ambika's calculation.

	H	T	O
	7	~~5~~ 4	15
–	4	2	7
	3	2	8

I wonder if I can check this answer with an addition.

755 – 427 = 328

so 328 = 755 – 427

Think together

1 One of these subtractions is incorrect.

Complete the part-whole models and choose additions to check the subtractions.

a) 612 – 371 = 341

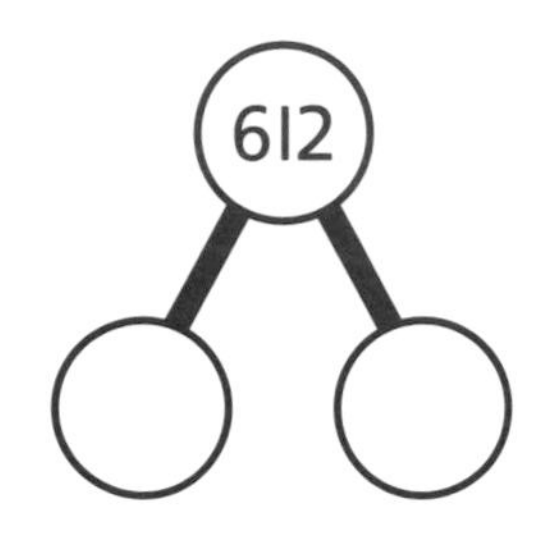

H	T	O
3	4	1
+		

b) 141 = 712 – 571

H	T	O
+		

Subtraction _______ is incorrect.

2 Use subtractions to check these additions.

a) 334 + 477 = 812

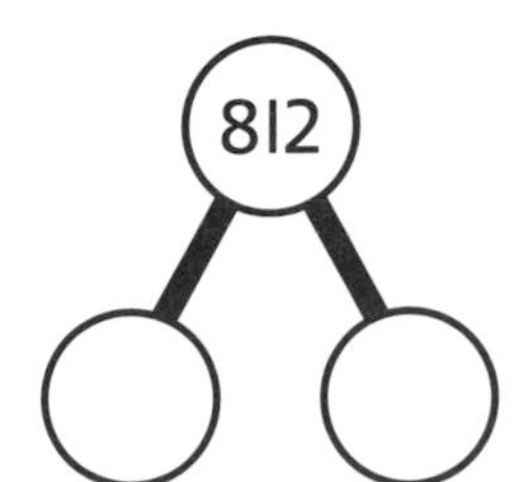

b) 812 = 521 + 391

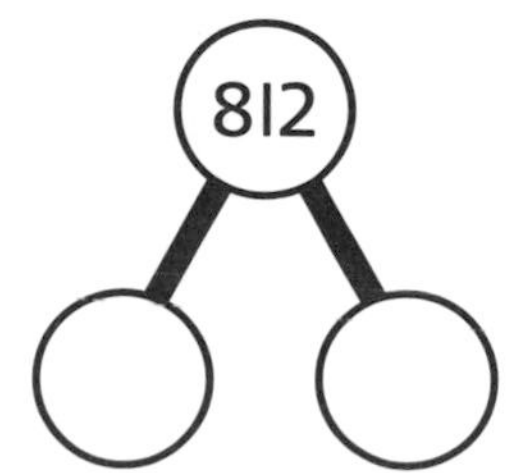

CHALLENGE

3 Emma has worked out 501 – 499 = 91

H	T	O
⁴~~5~~	0	¹⁰~~1~~
– 4	9	9
0	9	1

That doesn't look right. 499 is approximately 500, so the answer should be close to 0.

a) What two mistakes has she made?

b) Think about the fact family for 501 – 499 = ☐.

Which of these methods helps you to find the answer mentally?

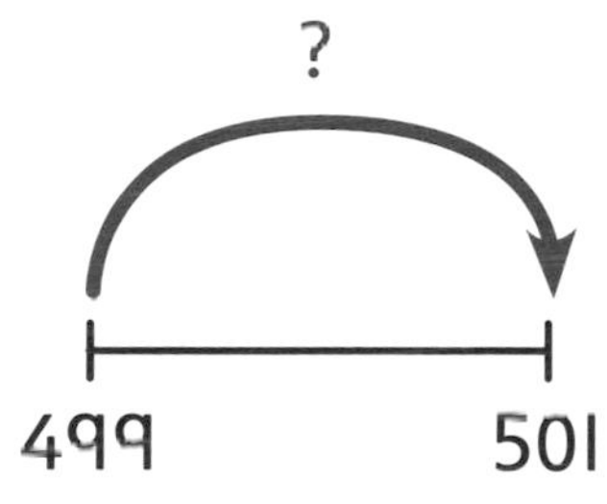
?
499
501

499 + ☐ = 501

4 Answer this calculation.

710 – ☐ = 690

I wonder if I can use a fact family to check 710 – ☐ = 690 mentally.

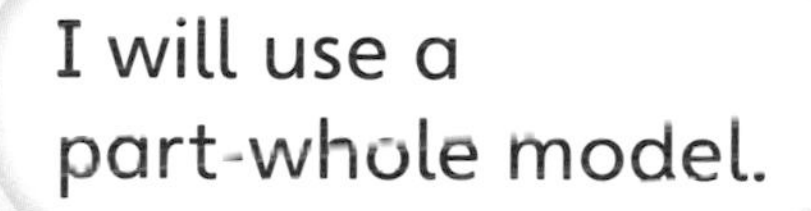
I will use a part-whole model.

→ Practice book 3A p92

Problem solving – addition and subtraction 1

Discover

1 a) Holly bought a racing bike and paid to have a service.

How much did she spend in total?

b) Write an approximation to check the calculation.

Share

a) A bar model shows the parts and the whole clearly.

I will start by drawing a bar model to help see what I need to work out.

?	
275	99

We know the parts but need to work out the whole.
I will add the parts together.

	H	T	O
	2	7	5
+		9	9
	3	7	4
		1	1

374	
275	99

Holly spent £374 in total.

b) £99 is approximately £100.

£275 + £100 = £375

£375 is very close to £374.

The answer looks correct.

I think I can approximate just one of the numbers.

Think together

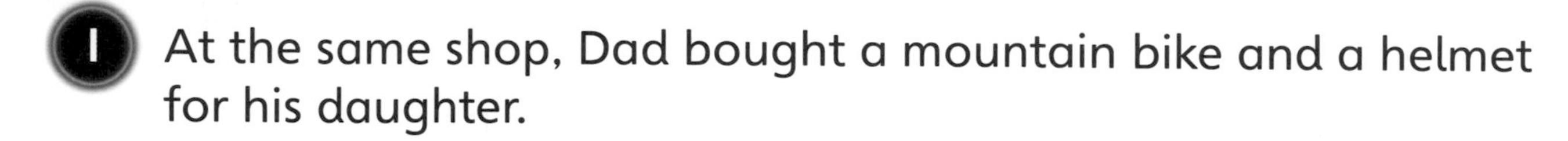

1 At the same shop, Dad bought a mountain bike and a helmet for his daughter.

How much did he spend?

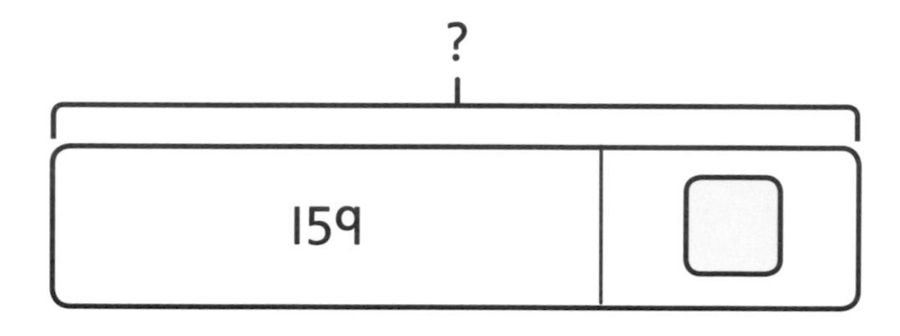

	H	T	O
	1	5	9
+			

Dad spent £[ ].

2 A family bought a tandem and a child's bike. They spent £468.

How much did the child's bike cost?

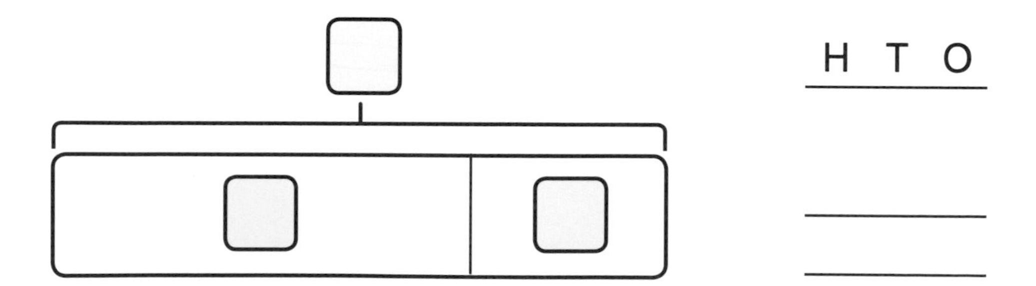

H	T	O

[ ] ◯ [ ] = [ ]

The cost of the child's bike was £[ ].

3 **a)** Toshi bought a racing bike, a helmet and lights.

How much did he spend altogether?

Draw models to show the steps in this problem.

I think there are two steps so I will try drawing two bar models.

There are three numbers to add, so I will draw one bar model with three parts.

b) Sofia bought a bike and two helmets. The total was £399.

Which bike did she buy?

What models can you use to show the steps of this problem?

Look at the picture of the bike shop to find the information you need.

→ Practice book 3A p95

Problem solving – addition and subtraction 2

Discover

1 a) How many more runs has Team A scored than Team B?

b) Bella and Andy start batting for Team B.

Bella scores 105 and Andy scores 83.

How many runs has Team B scored now?

Share

a) Team A has 454 runs. Team B has 128 runs.

I am comparing two numbers, so I will draw two bars.

Team A: 454

Team B: 128 ⟷ ?

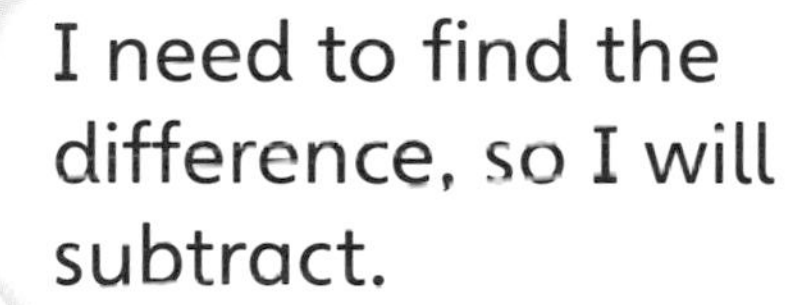
I need to find the difference, so I will subtract.

	H	T	O
	4	~~5~~ 4	[1]4
–	1	2	8
	3	2	6

454 – 128 = 326

Team A has scored 326 more runs than Team B.

b) I will add in two steps. First, I will add Bella's score. Then I will add Andy's score to the total.

128 + 105 = 233

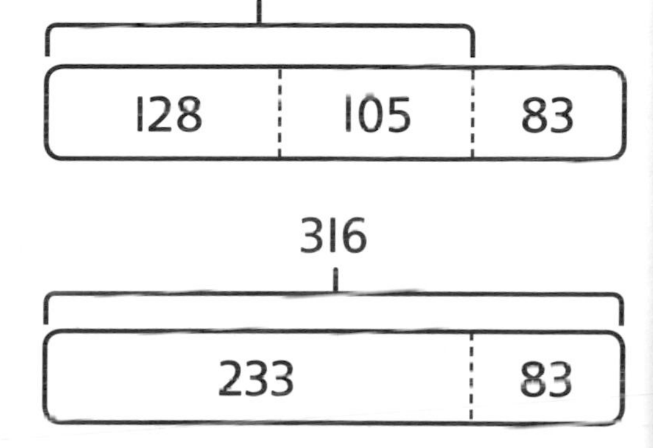
233
128
105
83
316
233
83

Team B has now scored 316 runs in total.

Think together

1 Aki's team scored 317 runs and Isla's team scored 451.

How many more runs did Isla's team score?

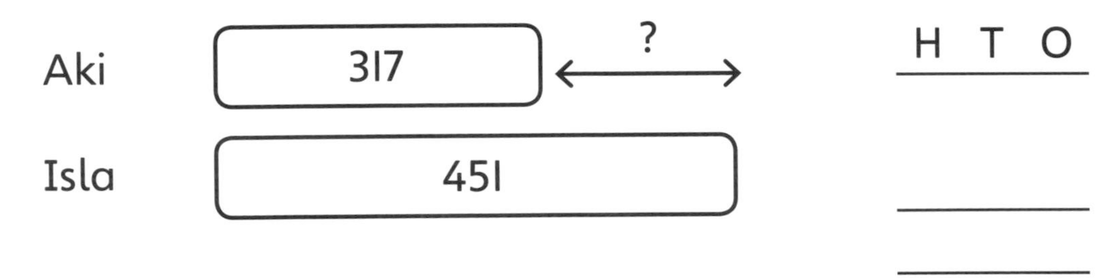

H	T	O

Isla's team scored ☐ more runs.

2 Mo and Lexi score 320 runs. Jamilla scores 165 and Emma scores 56.

How many more runs do they need in order to score the same as Mo and Lexi?

Mo and Lexi: 320

Jamilla and Emma: 165 | 56 ⟷

	H	T	O
	1	6	5
+		5	6

H	T	O

Jamilla and Emma need ☐ more runs to get the same score as Mo and Lexi.

3 Richard scores 188 runs and Olivia scores 56 more than Richard. How many runs do they score altogether?

I made a bar with three parts.

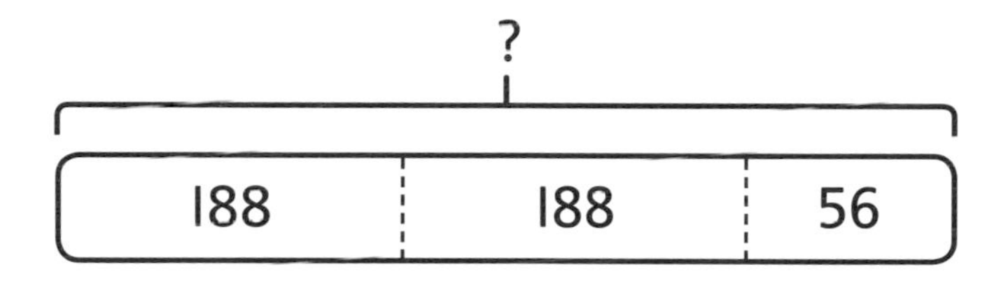

I made two bars to show that Olivia scored more than Richard.

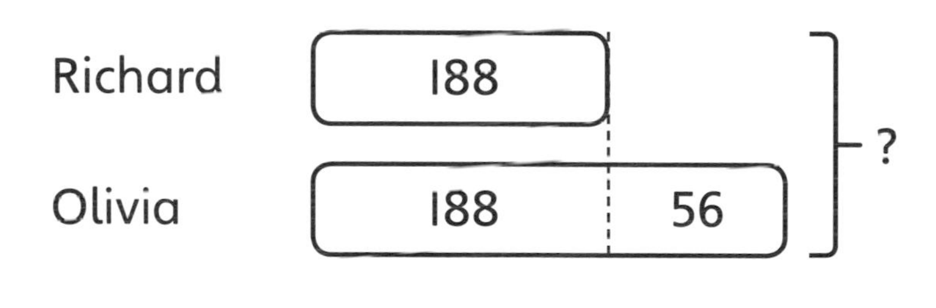

Which bar model shows the problem better?

Copy the bar model and then solve the problem.

→ Practice book 3A p98

End of unit check

1 What is missing from this calculation?

333 ☐ = 353

A + 2 **B** + 200 **C** + 20 **D** – 20

2 Which addition exchanges 10 ones **and** 10 tens?

A

	H	T	O
	2	0	1
+	3	0	9

B

	H	T	O
	4	1	0
+	3	9	0

C

	H	T	O
	4	2	2
+	3	9	7

D

	H	T	O
	4	1	2
+	3	8	9

3 Which subtraction is not correct?

A

	H	T	O
	$^{2}\not{3}$	$^{1}5$	0
–	1	8	0
	1	7	0

B

	H	T	O
	$^{2}\not{3}$	$^{1}5$	4
–	1	8	5
	1	7	1

C

	H	T	O
	$^{2}\not{3}$	$^{1}5$	5
–	1	8	4
	1	7	1

D

	H	T	O
	3	8	5
–	2	1	5
	1	7	0

4 Which calculation has an answer that is approximately 500?

A 901 – 399 **B** 401 + 198 **C** 350 + 248 **D** 999 – 598

5 Richard has done an addition to check his calculation.

325 + 476 =

Which problem was he trying to solve?

A 476 – 325

B 325 – 476

C 801 – 325

D 801 – 376

6 Tim and Alanna each have a length of wool. Tim's is 500 cm long. Alanna cuts 175 cm off her wool. Now it is the same length as Tim's.

How long was Alanna's wool to start with?

→ Practice book 3A p101

Unit 4
Multiplication and division 1

In this unit we will ...

- Recognise when groups are equal and when they are not
- Learn the 3, 4 and 8 times-tables
- Find a simple remainder when a number is divided
- Use a bar model to solve multiplication and division problems

In Year 2, we recognised when groups were equal and unequal.

Equal groups

Unequal groups

We will need some maths words. How many of these have you used before?

equal **multiply** **divide**

times-tables **sharing** **grouping**

array **bar model** **remainder**

repeated addition **multiplication sentence**

division statement **division facts**

You need to know that an array can tell you two different multiplication facts.

5 groups of 2

5 × 2 = 10

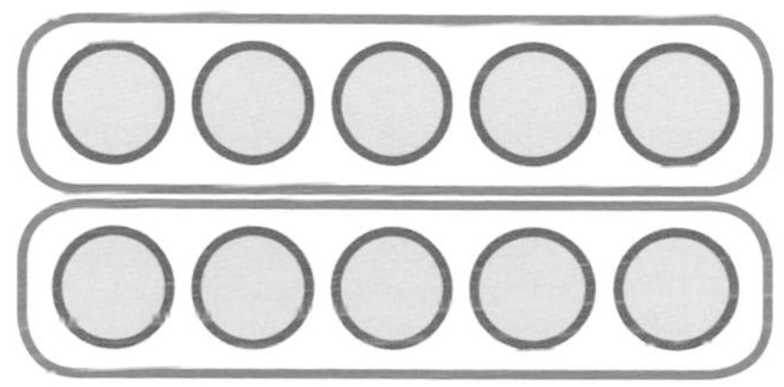

2 groups of 5

2 × 5 = 10

Multiplication – equal grouping

Discover

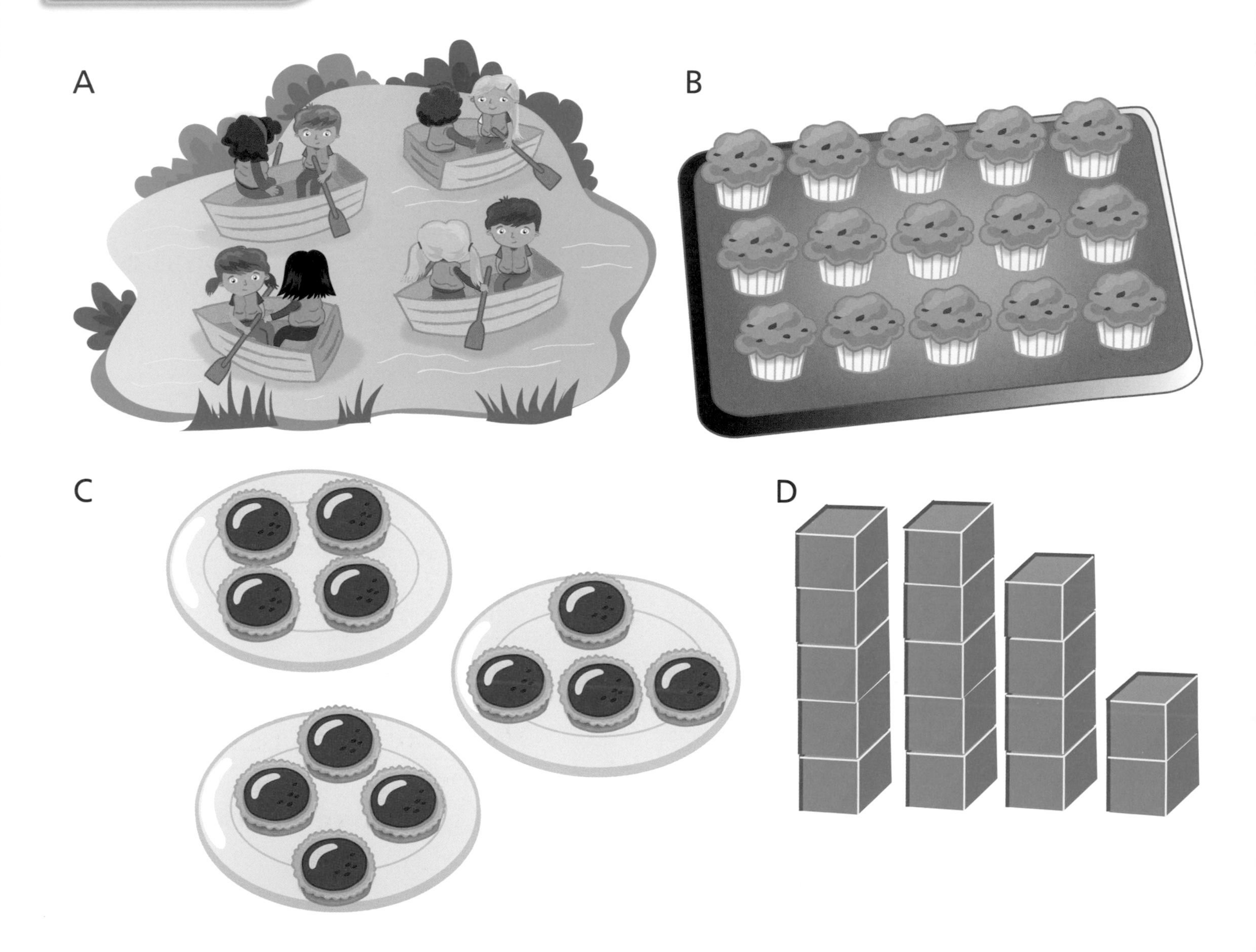

1 **a)** Which images show equal groups?

How do you know?

b) For each equal group, complete a **multiplication sentence**.

Share

a) A

There are 4 equal groups of 2 people.

B

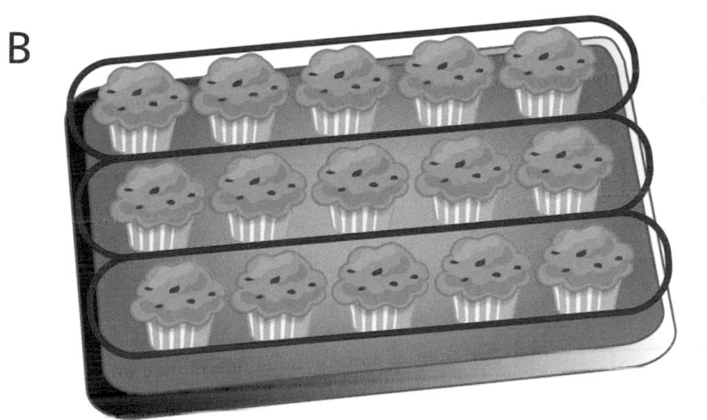

There are 3 equal groups of 5 muffins.

C

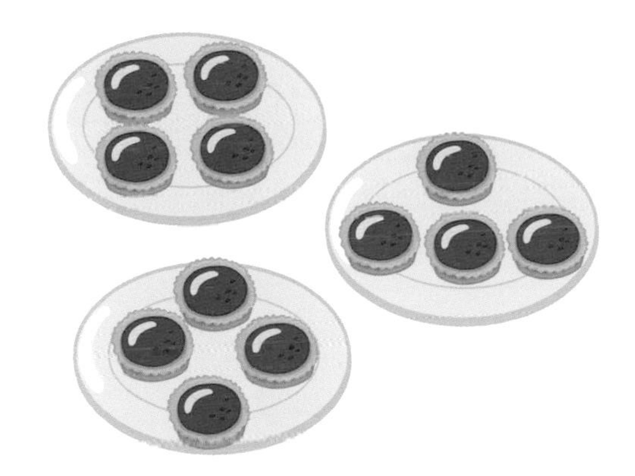

There are 3 equal groups of 4 tarts.

I worked out 3 × 4 as 4 + 4 + 4 = 12. I remembered that multiplication is the same as **repeated addition**.

D

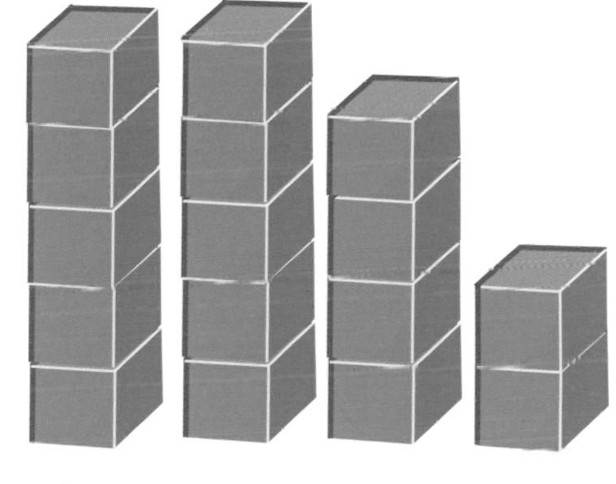

These are not equal groups.

There are 2 groups of 5 cubes, 1 group of 4 cubes and 1 group of 2 cubes.

b) The total number of people is 4 × 2 = 8.

The total number of muffins is 3 × 5 = 15.

The total number of tarts is 3 × 4 = 12.

Think together

1 How many counters are there in total?

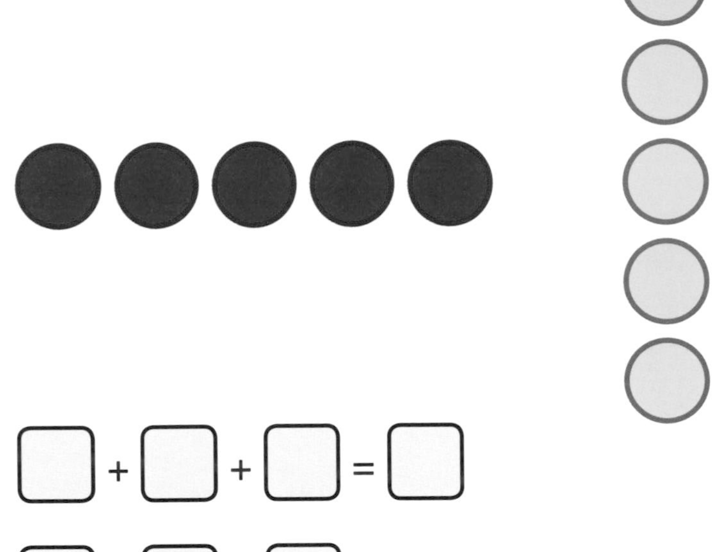

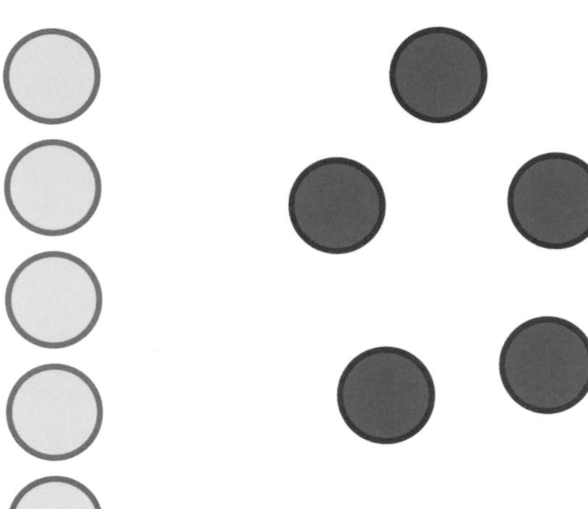

2 How many counters are there in total?

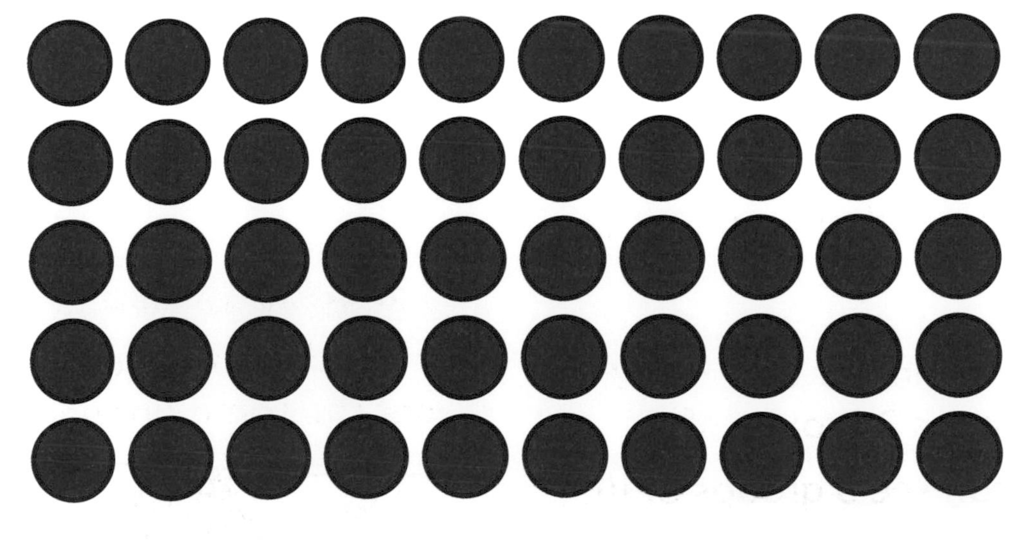

I wonder if there is more than one multiplication to find the total.

CHALLENGE

3 Ebo, Alex and Aki are working out how many stars there are in total.

I did 5 + 5 + 5 + 5 + 5 + 5
Aki
I did 3 × 10
I did 15 × 2
Ebo
Alex

For each person, circle the groups with your finger.

Can you see any other equal groups?

I remember that 3 × 10 means 3 groups of 10.
I can see 3 groups of 10.

I wonder if they all get the same answer.

→ Practice book 3A p103

Multiplying by 3

Discover

1 **a)** There are 3 balls under each cup.

How many balls are there in total?

Write down a multiplication statement to work out the answer.

b) Work out 8×3.

Share

a) Under each cup there are 3 balls.

I counted up in 3s, using a number line to help me.

I could count them one by one.

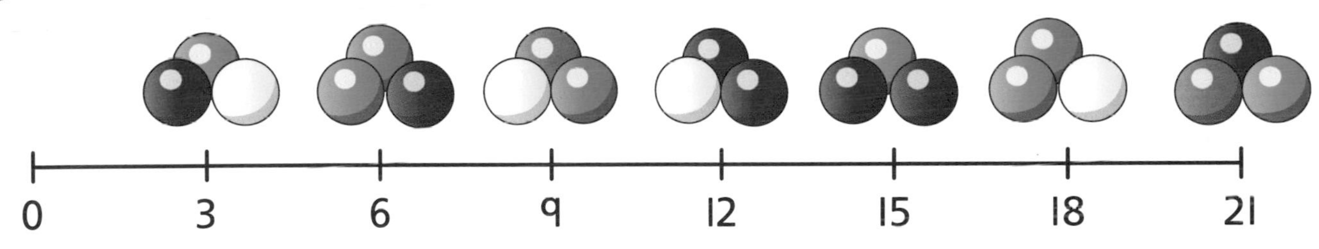

3 + 3 + 3 + 3 + 3 + 3 + 3 = 21

7 × 3 = 21

There are 21 balls in total.

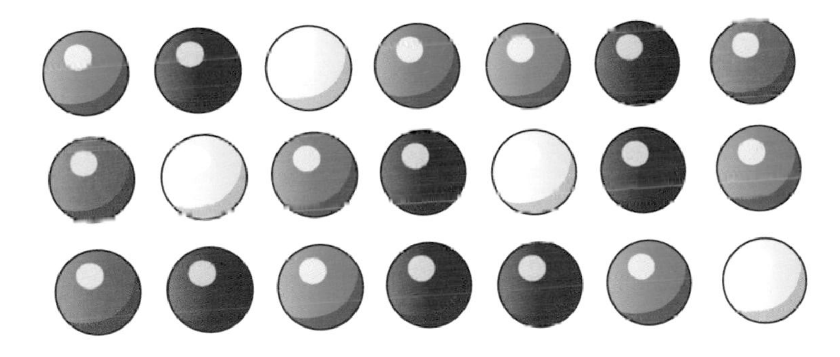

Now I know 7 groups of 3, I can easily work out 8 groups of 3.

This is a 7 × 3 array. 7 × 3 = 21

b) 8 × 3 = 24

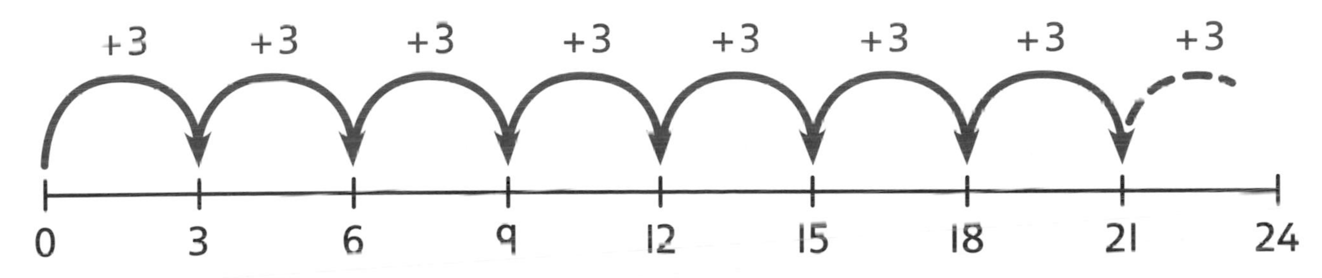

Think together

1 There are 3 balls under each cup.

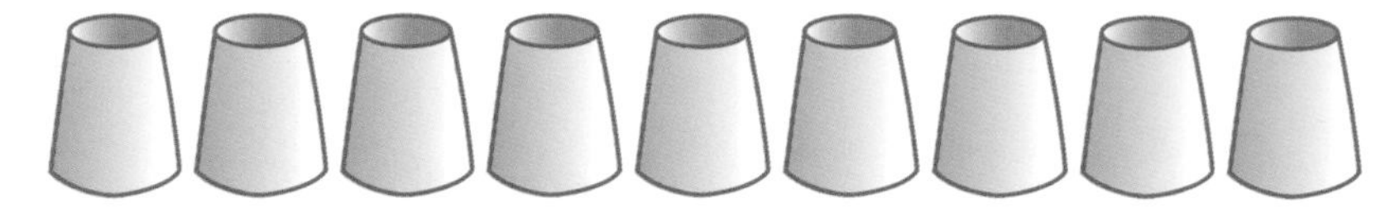

How many balls are there?

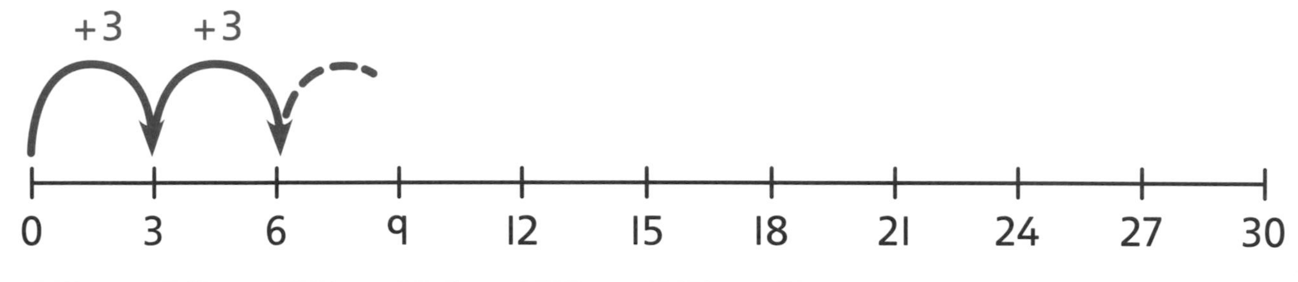

□ + □ + □ + □ + □ + □ + □ + □ + □ = □

□ × 3 = □

There are □ balls.

2 How many hats are there?

□ × □ = □

There are □ hats.

3 What is the same? What is different?

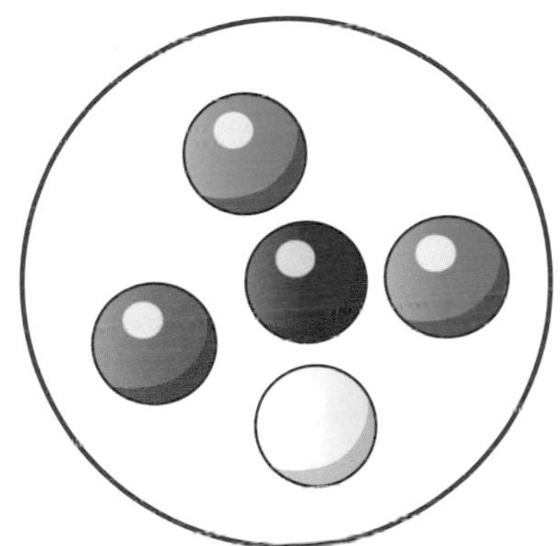

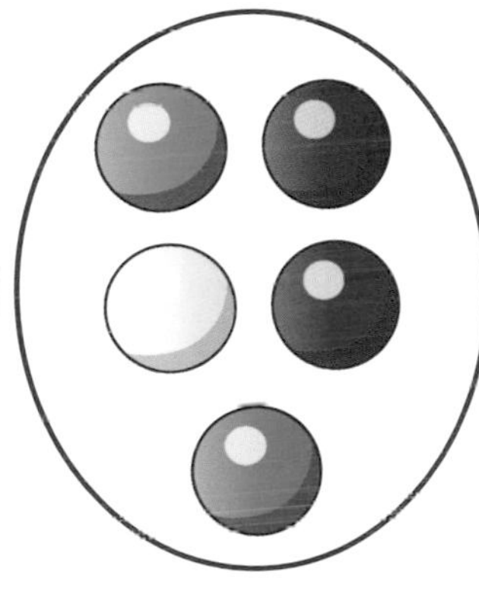

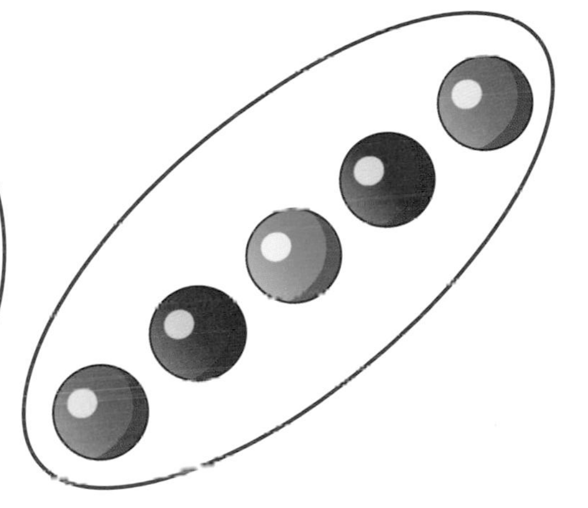

Discuss with your friend.

I think they all have the same number of objects.

There are 3 groups of 5 marbles. I wonder if they could make equal groups another way.

→ Practice book 3A p106

Dividing by 3

Discover

1 **a)** Each box holds 3 cupcakes.

How many boxes are needed for all the cupcakes?

Work this out by writing a **division statement**. 

b) David buys 27 cakes. He shares them equally between 3 people.

How many cakes do they get each?

Share

If I put 3 cupcakes into each box then I have 3 fewer each time. I will use a number line to jump back.

a) There are 18 cupcakes.

Each box holds 3 cupcakes.

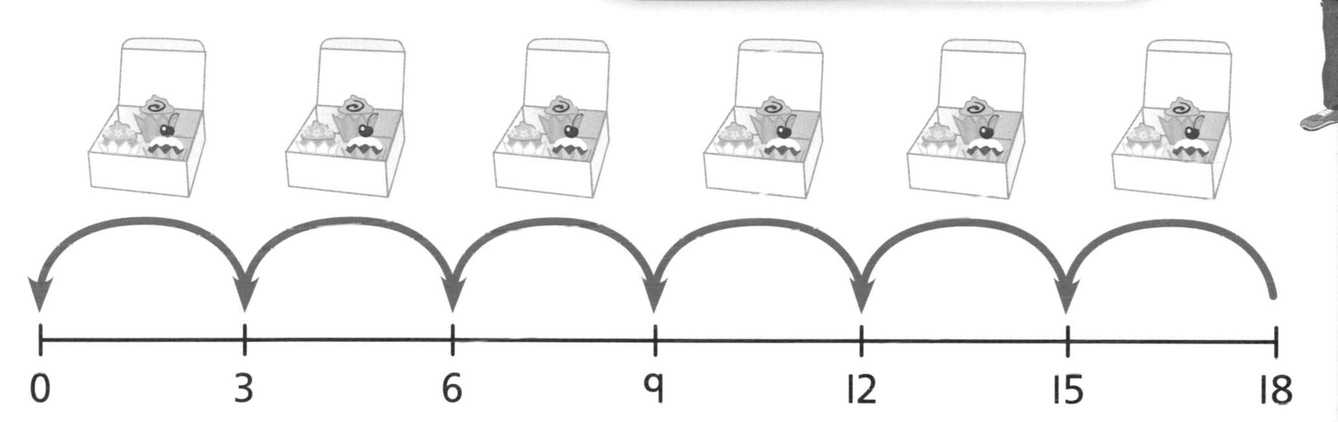

18 ÷ 3 = 6

6 boxes are needed.

I shared them out 1 at a time. I could have shared them 3 at a time.

b)

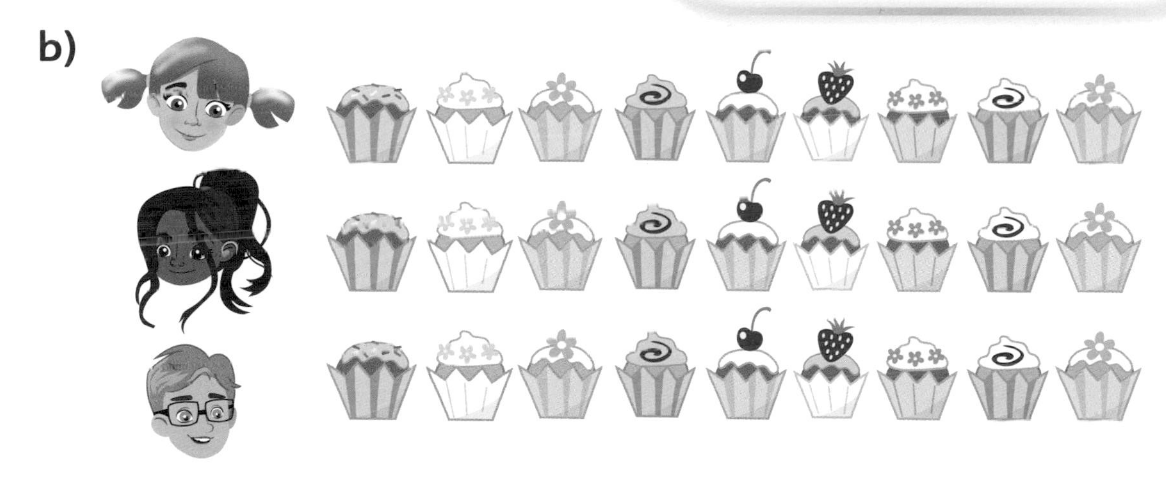

There are 27 cakes.

There are 3 people.

Each person gets 9 cakes.

27 ÷ 3 = 9

Think together

1 Bread rolls are packed in 3s.

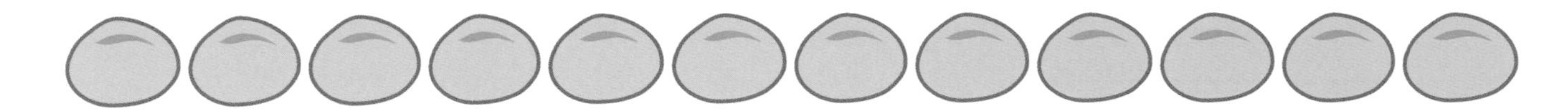

How many packs can be made?

There are □ bread rolls.

There are 3 bread rolls in each pack.

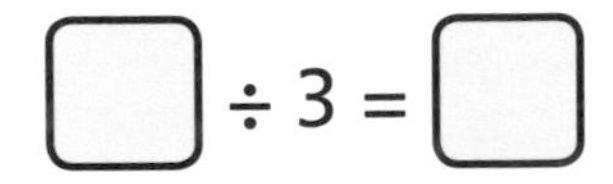

So □ packs can be made.

2 21 doughnuts are shared equally between 3 plates.

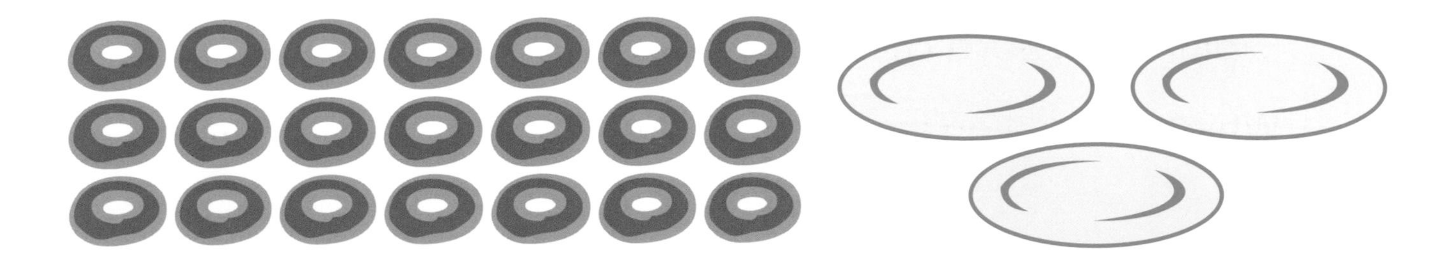

How many doughnuts will go on each plate?

□ ÷ □ = □

So □ doughnuts will go on each plate.

3 Class 3A have been set a question.

$$33 \div 3$$

Here are three methods.

Lee

I drew an array to help me.

Zac

I took 33 counters and put them into groups of 3.

Olivia

I am going to use a multiplication fact that I know off by heart to get the answer.

Explain how each child got the answer.

Which method do you prefer? What other method could you use?

I prefer Zac's method because you have to make groups of 3 for division.

I wonder if you could also use sharing.

→ Practice book 3A p109

3 times-table

Discover

1 a) What are the missing answers?

How did you work them out?

b) Which multiplication fact can help you work out the total?

Share

I will draw each as an array to help me work them out ...

a) First look at the multiplications.

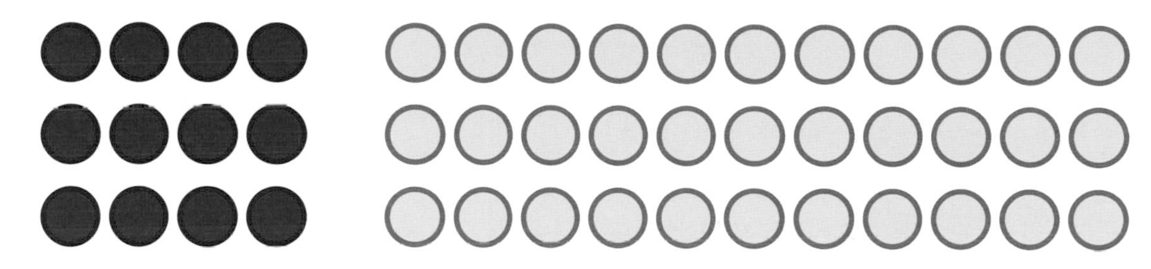

4 × 3 = 12

11 × 3 = 33

... I grouped to find the answers to the division questions.

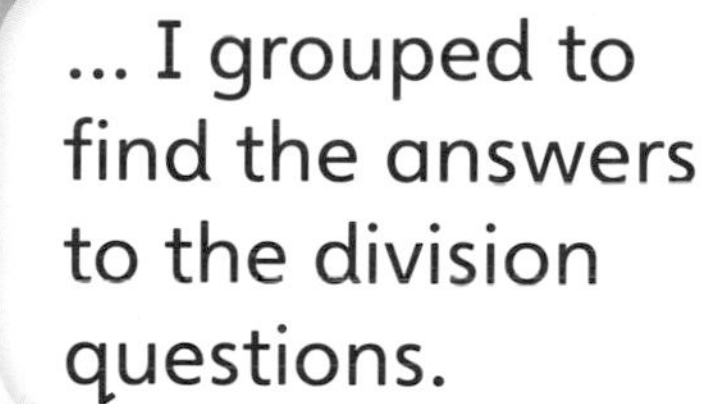

Now look at the divisions.

9 ÷ 3 = 3

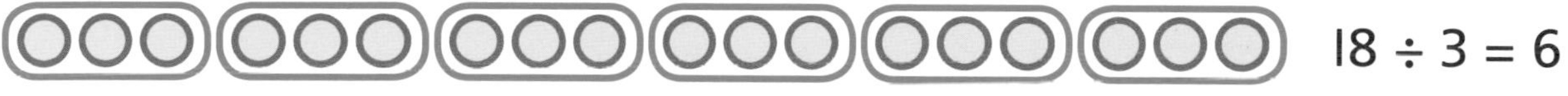

18 ÷ 3 = 6

You can use multiplication facts to help you. If 6 × 3 = 18, then 18 ÷ 3 = 6. Can you see the link? It is important to know **times-table** facts.

b)

5 × 3 = 15

Think together

1 Use multiplication facts to work out how many items there are in each picture.

Which fact did you use to find each total?

a) 

b) 

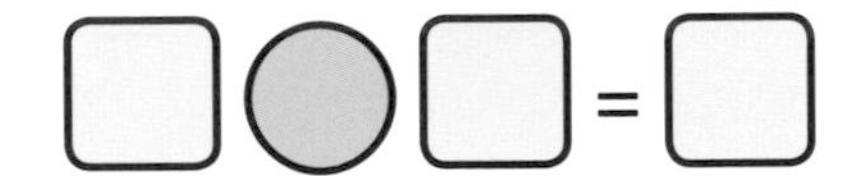

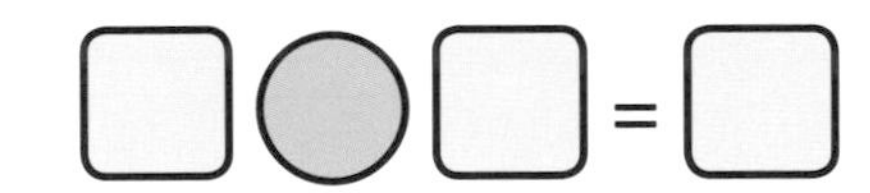

2 How many of these can you work out in a minute?

11 × 3	9 × 3	0 × 3	12 ÷ 3
3 × 7	8 multiplied by 3	☐ × 3 = 36	multiply 2 by 3
divide 30 by 3	3 × ☐ = 18	24 ÷ 3	number of questions in this grid

It is not always important to be quick, but knowing your times-table facts can help save time.

a) Find two different methods for Luis to use to work out 12×3.

Which method do you prefer?

Which method is quicker?

b) How can you work these out using the 3 times-table?

$3 \times 3 \times 3$ 13×3 3×20

If Emma already knows $6 \times 3 = 18$, then I think she can work it out.

I wonder how many 3 times-table facts I know off by heart. I will cover them up and see if I can remember them.

→ Practice book 3A p112

Multiplying by 4

Discover

1 **a)** There are 6 donkeys.

How many donkey legs are there in total?

Write a multiplication statement to work out the answer.

b) A family of 5 people are going donkey trekking.

Mr Peters pays 20 £1 coins in total for him and his family.

Is this the correct amount?

Share

a) There are 6 donkeys.

Each donkey has 4 legs.

Instead of counting the legs in 1s, I am going to count up in 4s.

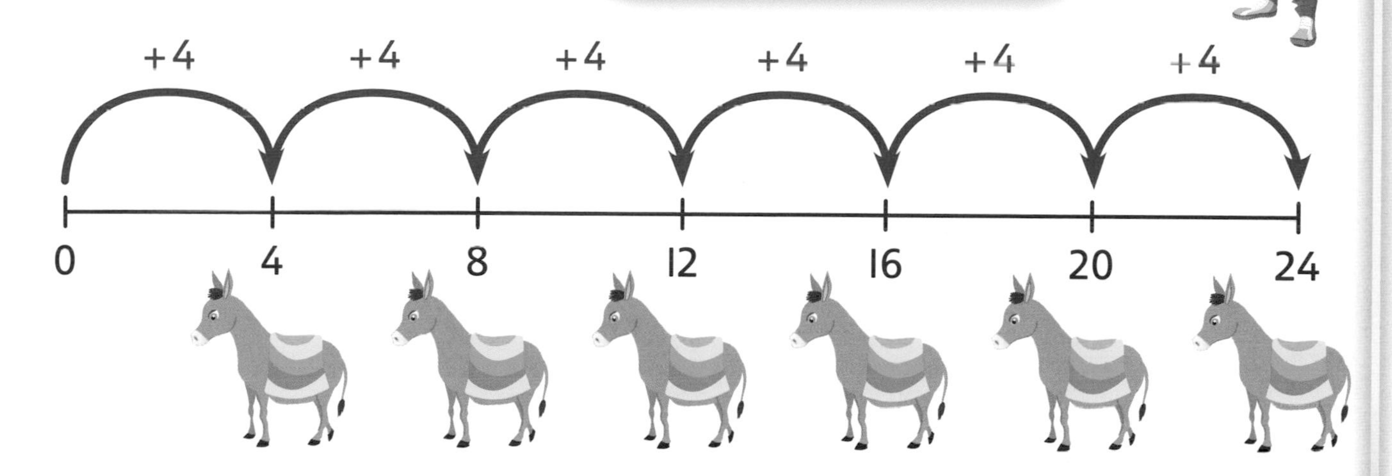

$6 \times 4 = 24$

There are 24 donkey legs in total.

Remember, we can think of 6×4 as meaning 6 groups of 4, which is what we have.

b) There are 5 people in the family.

The cost for each person is 4 £1 coins.

The total cost is $5 \times 4 = 20$ £1 coins.

Mr Peters pays the correct amount.

Think together

a) There are 7 donkeys.

How many donkey legs are there in total?

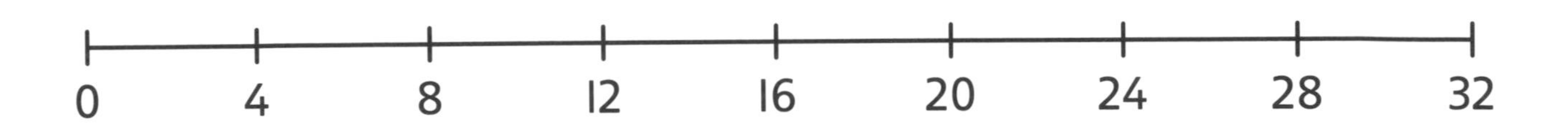

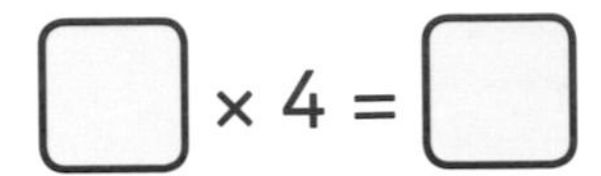

There are ☐ donkey legs.

b) How much does it cost in total for 4 people to go donkey trekking?

It costs ☐ £1 coins in total to go donkey trekking.

I wonder if I can use my answers from earlier.

2 There are 32 donkey legs. How many donkeys are there?

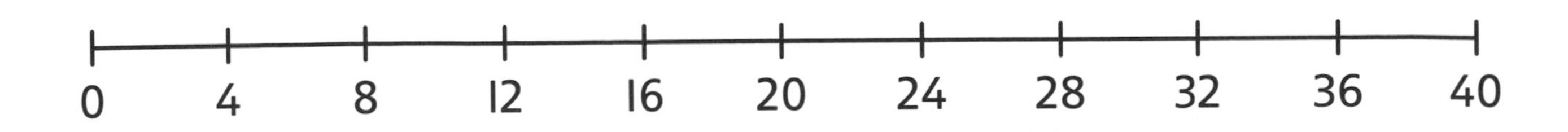

There are ☐ donkeys because ☐ × ☐ = ☐.

CHALLENGE

3 A box contains 4 apples.

How many apples are there in 9 boxes?

To work this out you have to do 9 × 4.

I have a different way of multiplying by 4. You can multiply by 2 and then multiply by 2 again.

Ebo did the following working:

$9 \times 2 = 18$
$18 \times 2 = 36$
So, $9 \times 4 = 36$

Does this work for 10×4?

What about 6×4?

Check it with your own numbers.

That's interesting! To multiply by 4, I can double and then double again!

Use equipment to show why this works.

I can show this using a picture or cubes.

→ Practice book 3A p115

Dividing by 4

Discover

1 **a)** 20 cards are dealt equally between 4 players.

How many cards does each player get?

Write this as a division statement.

b) The left-over cards are put into piles of 4.

How many piles are formed?

Share

I will give the cards out 1 at a time.

a) 20 cards are dealt out.

There are 4 players.

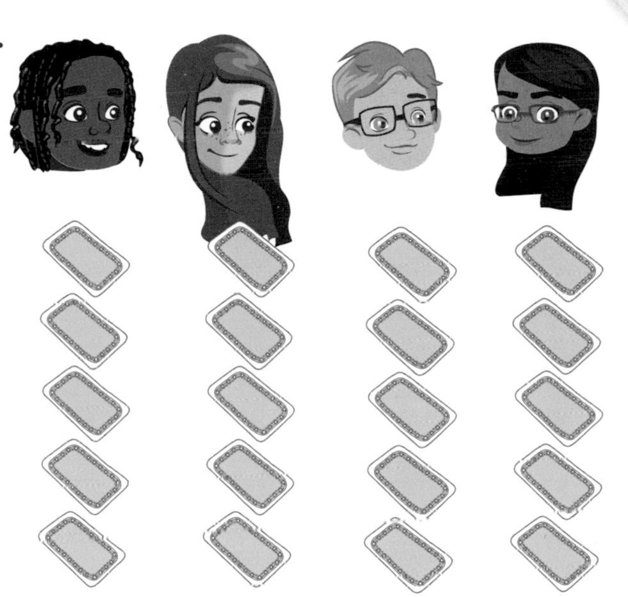

20 ÷ 4 = 5

Each player gets 5 cards.

b) There are 48 cards in the box in total.

20 cards have been dealt already.

There are 28 cards left in the pack.

I grouped them into 4s until there were none left.

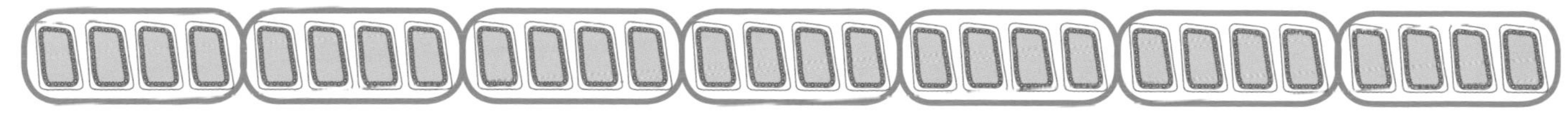

0 1 2 3 4 5 6 7 8 9 10 11 12 13 14 15 16 17 18 19 20 21 22 23 24 25 26 27 28

There are 28 cards.

There are 4 cards in each group.

28 ÷ 4 = 7

7 piles are formed.

Think together

1 12 coins are shared between 4 money boxes.

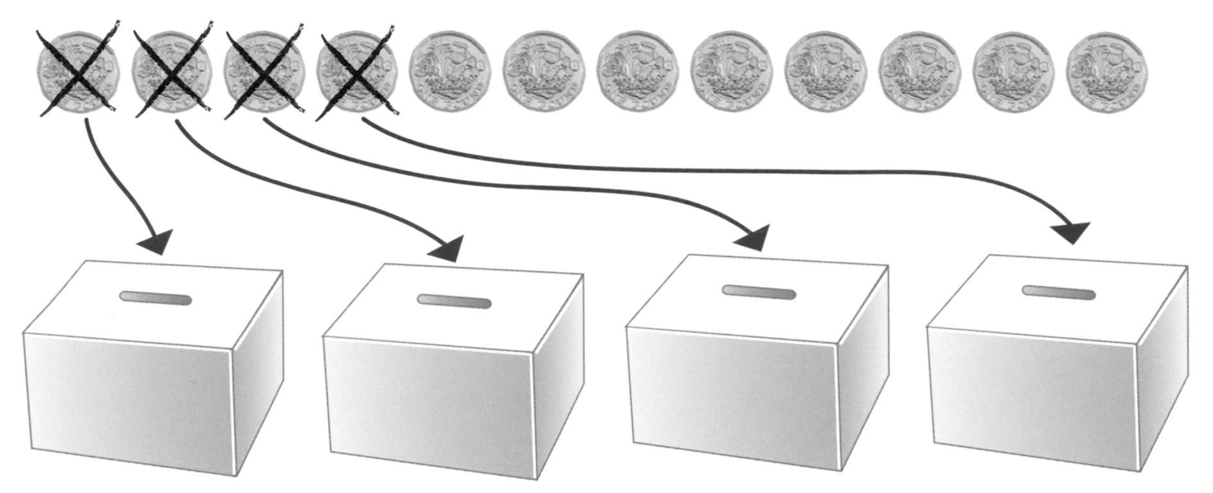

There are ☐ coins.

There are 4 money boxes.

☐ ÷ 4 = ☐

There are ☐ coins in each money box.

2 There are 32 flowers.

The flowers are put into bunches of 4.

How many bunches are there?

☐ ÷ ☐ = ☐

There are ☐ bunches.

3 a) 44 marbles are shared equally between the boxes.

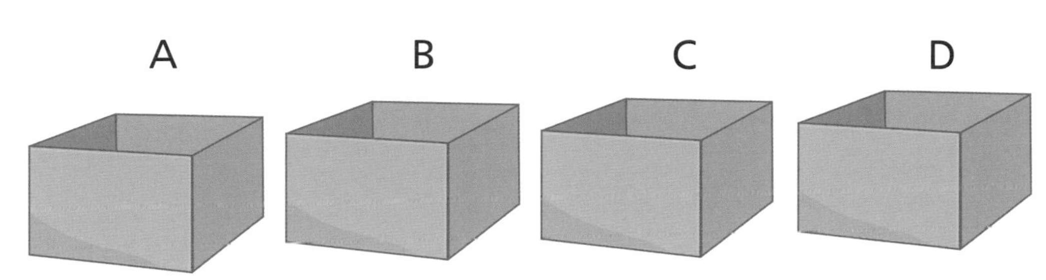

I have 44 marbles. I know I need to divide by 4. I can divide by 2 and by 2 again. This will tell me how many I should put in each box.
Jamie

How many marbles are there in each box?

b) What is 22 marbles shared equally between 4 boxes?

I think Jamie's method works every time.

I wonder which numbers divide by 4 and which do not. Is there any way of telling?

→ Practice book 3A p118

4 times-table

Discover

1 **a)** What numbers are the children covering?

How did you work them out?

b) Which multiplication facts will help you work out these calculations?

Work out the answers.

$4 \times 7 = \square$ $48 \div 4 = \square$

Share

a) Create arrays using counters to find the answers.

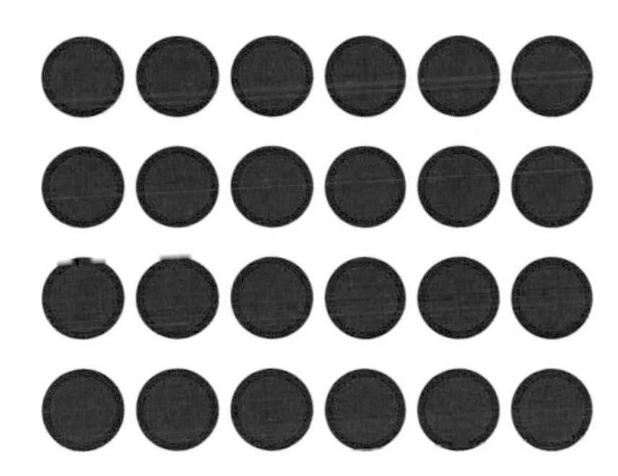

$6 \times 4 = 24$

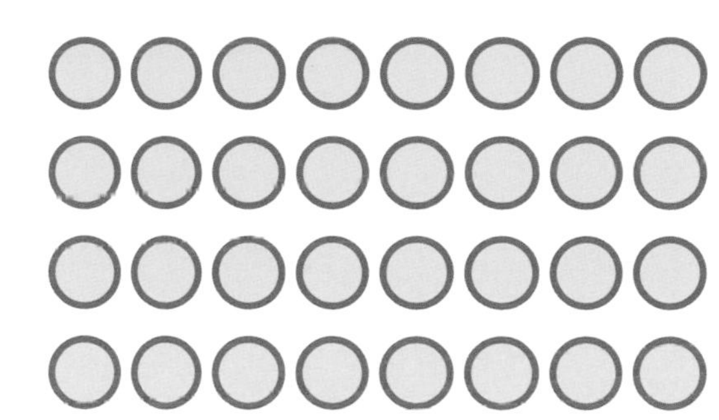

$8 \times 4 = 32$

The children are covering 24 and 32.

b)

I know that 4×7 is the same as 7×4.

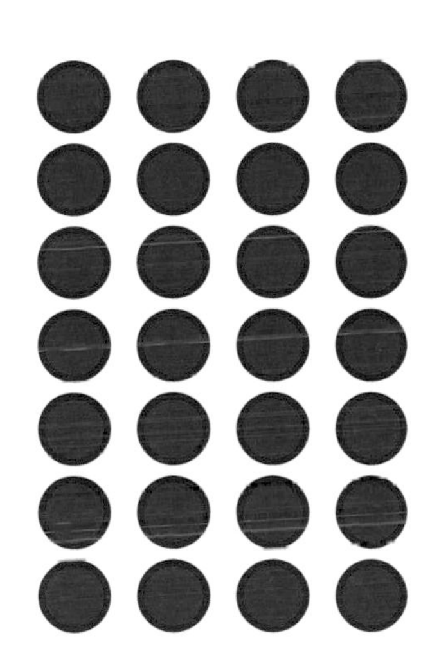

$4 \times 7 = 28$

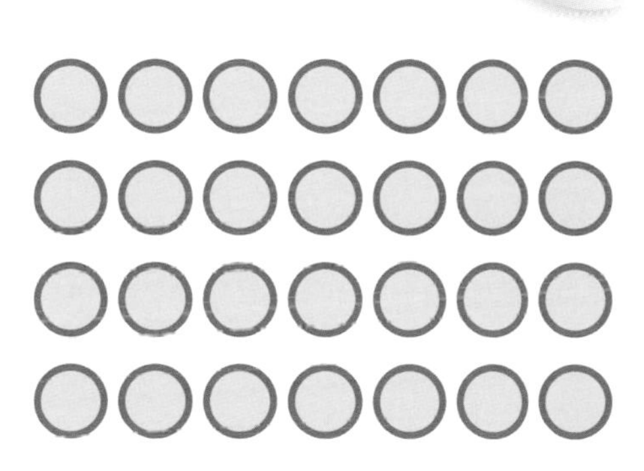

$7 \times 4 = 28$

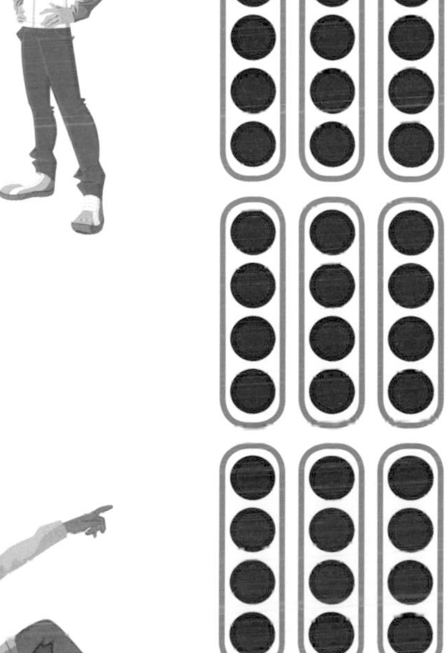

$48 \div 4 = 12$

If I know $12 \times 4 = 48$,
I also know $48 \div 4 = 12$.

Think together

1 Use multiplication facts to work out how many there are of each item in total.

a)

There are ☐ cubes.

b)

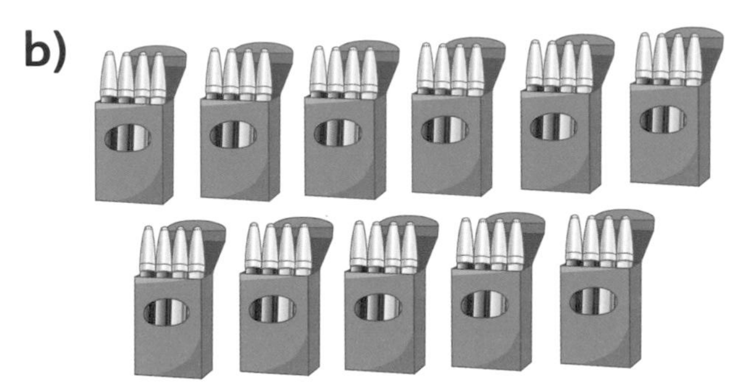

There are ☐ boxed pens.

c)

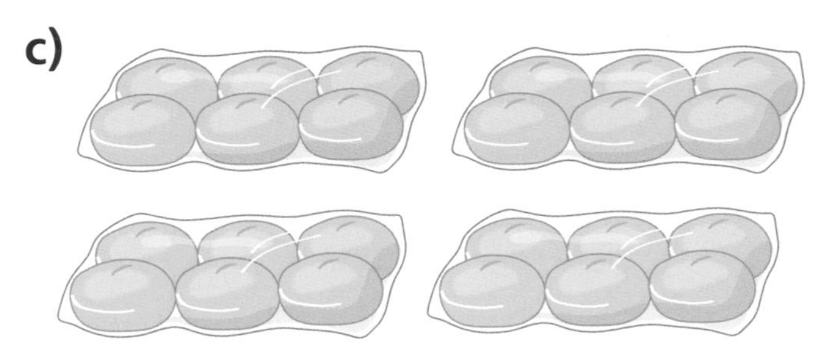

There are ☐ bread rolls.

Which fact did you use each time?

2 Mary has been asked some questions.

How many has Mary got right?

7 × 4 = 28	12 ÷ 4 = 3
4 × 9 = 36	4 ÷ 4 = 0
4 × 1 = 4	8 ÷ 4 = 32
0 × 4 = 4	24 ÷ 4 = 8
10 × 4 = 43	44 ÷ 4 = 11

Mary has got ☐ answers right.

I can tell straightaway that one of the multiplication answers is wrong!

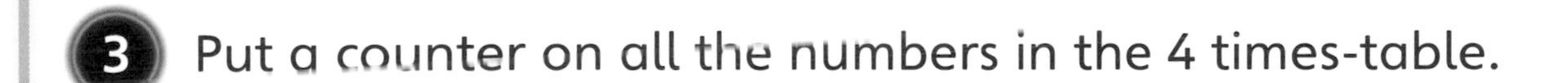

3 Put a counter on all the numbers in the 4 times-table.

1	2	3	4	5	6	7	8	9	10
11	12	13	14	15	16	17	18	19	20
21	22	23	24	25	26	27	28	29	30
31	32	33	34	35	36	37	38	39	40
41	42	43	44	45	46	47	48	49	50
51	52	53	54	55	56	57	58	59	60
61	62	63	64	65	66	67	68	69	70
71	72	73	74	75	76	77	78	79	80
81	82	83	84	85	86	87	88	89	90
91	92	93	94	95	96	97	98	99	100

You could go on finding numbers even bigger than 100 that are in the 4 times-table.

What patterns do you notice?

What do the numbers have in common?

I wonder how the 4 times-table relates to the 2 times-table. I will use a different colour for the 2 times-table.

All the numbers so far are even. I wonder if that is true for any number in the 4 times-table.

→ Practice book 3A p121

Multiplying by 8

Discover

1 **a)** Each pie has been cut into 8 slices.

How many slices are there in total?

Write down a multiplication statement to work out the answer.

b) $5 \times 2 = \square$ $5 \times 4 = \square$ $5 \times 8 = \square$

What do you notice?

Share

a) Each pie is cut into 8 slices.

There are 4 pies.

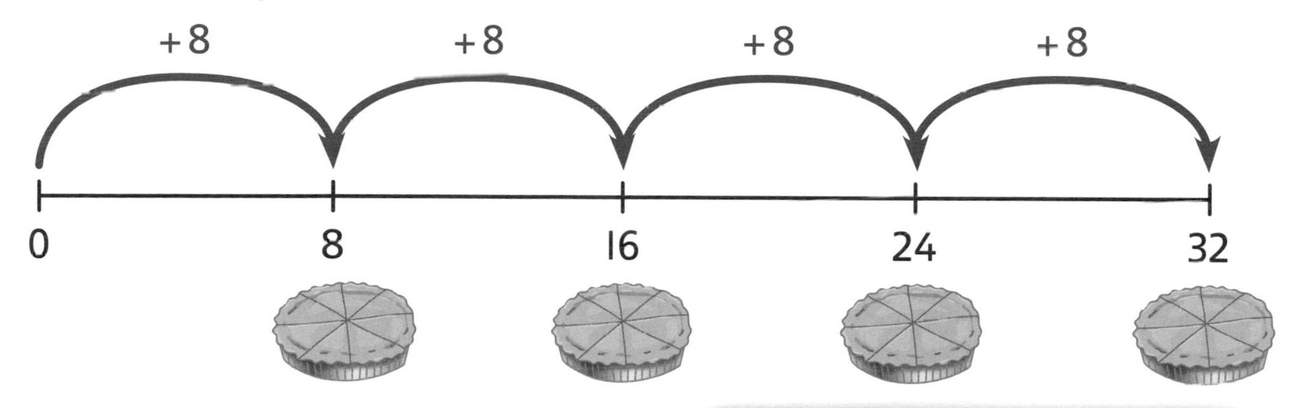

4 × 8 = 32

There are 32 slices in total.

Remember, we can think of 4 × 8 as meaning 4 groups of 8, which is what we have.

b)

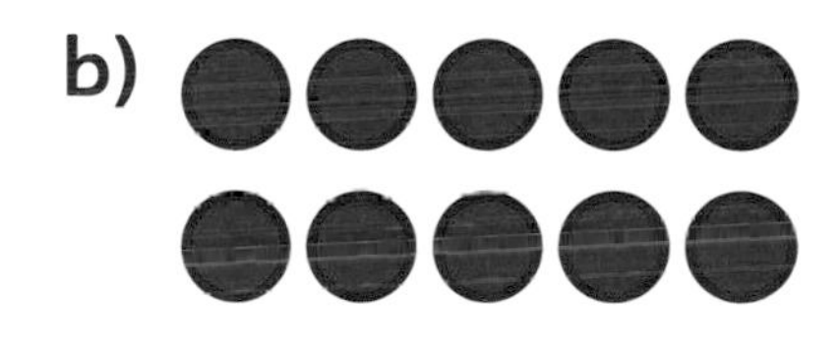

5 × 2 = 10

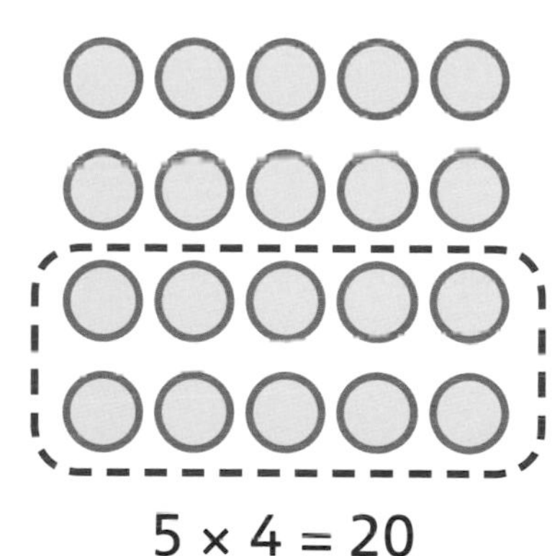

5 × 4 = 20

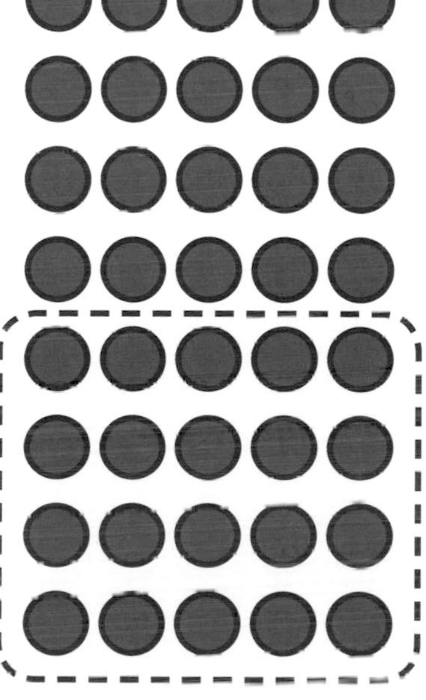

5 × 8 = 40

The answer doubles each time. I wonder if this works if I change the 5 to any other number.

Think together

1 A spider has 8 legs.

How many legs are there altogether?

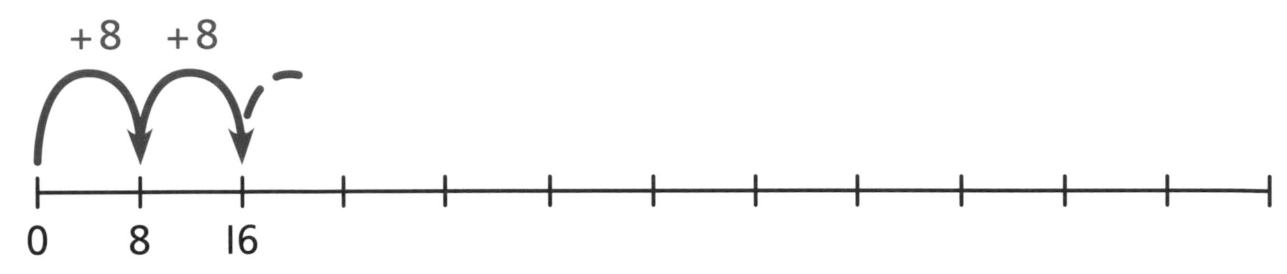

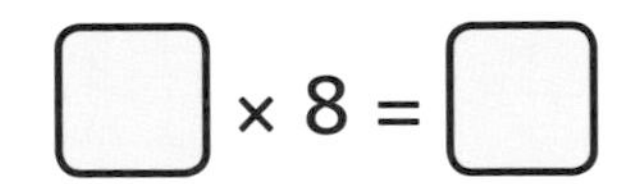

There are ☐ legs altogether.

2 A ticket to see a play is £8.

11 people are waiting to buy tickets.

How much will it cost them in total?

☐ × ☐ = ☐

The total cost is £☐.

3 Here are two methods for multiplying.

Here is my method for multiplying by 4.

Here is my method for multiplying by 8.

To multiply by 4

First, double your number (multiply by 2).

Then double your answer (multiply by 2 again).

To multiply by 8

First, double your number (multiply by 2).

Then double your answer (multiply by 2 again).

Then double your answer again (multiply by 2 again).

a) Use Luis's method to work out 9 × 4.

b) Use Isla's method to work out 9 × 8.

c) Work out 15 × 4, and 15 × 8.

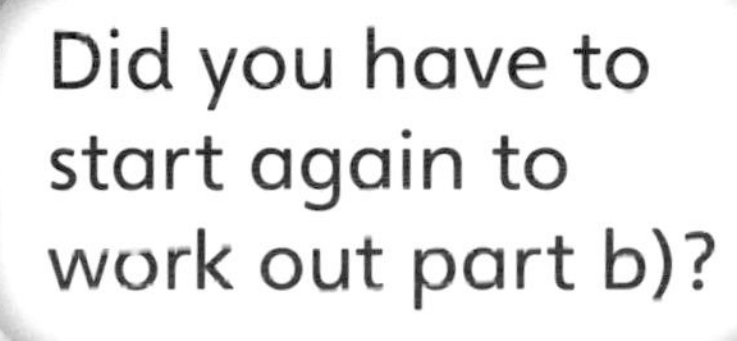

Did you have to start again to work out part b)?

I wonder if this works for any number.

If so, I wonder why it works.

→ Practice book 3A p124

Dividing by 8

Discover

1 **a)** Each ice lolly mould uses 8 lollipop sticks.

Mr Jones has 24 sticks.

How many moulds can he fill?

b) Miss Hall has 38 sticks.

How many moulds can she fill?

Share

a) Mr Jones has 24 lollipop sticks.

Each mould uses 8 sticks.

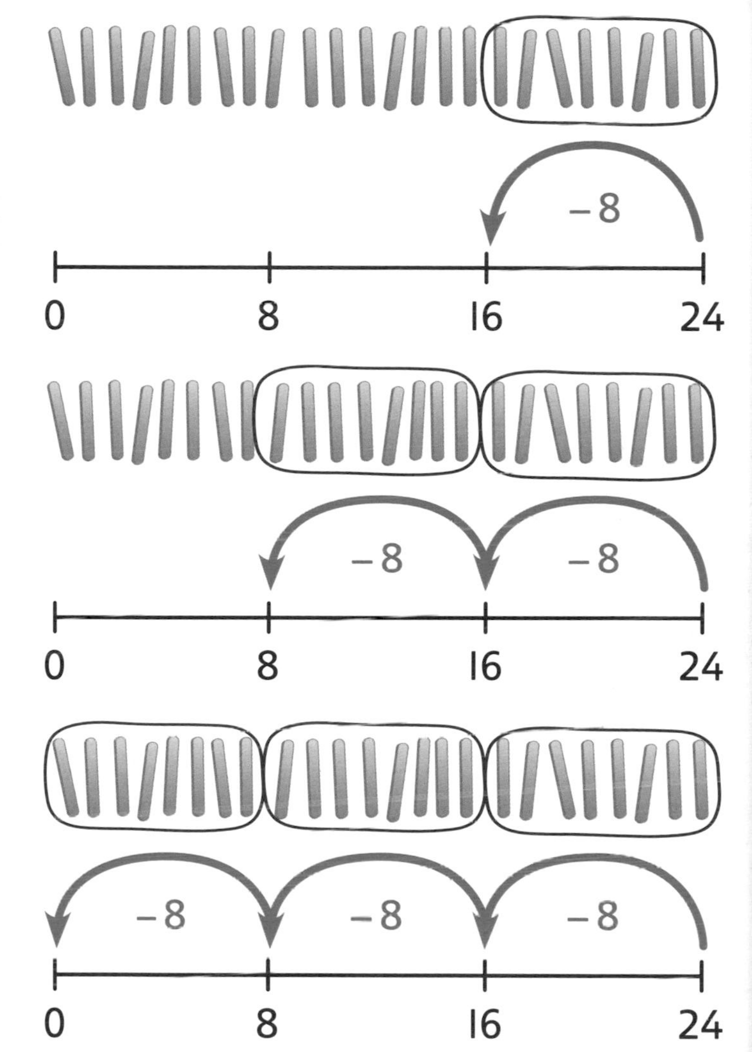

24 ÷ 8 = 3

Mr Jones can fill 3 moulds.

b) Miss Hall has 38 sticks.

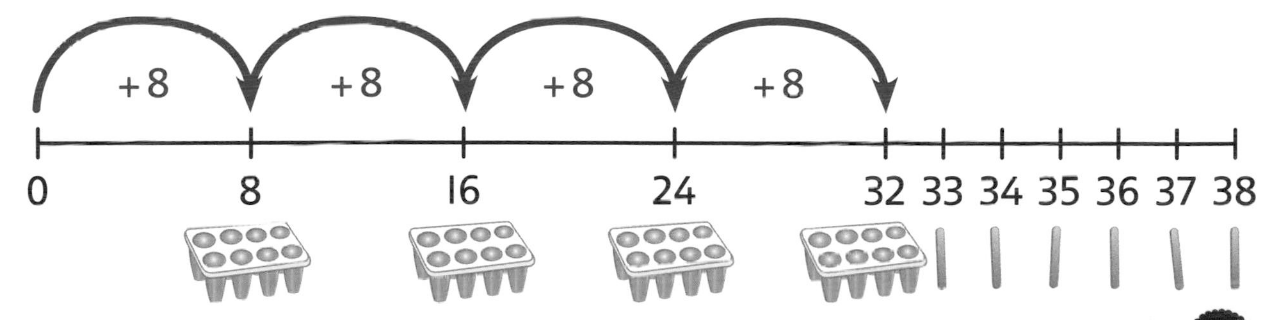

Miss Hall can fill 4 moulds.

There aren't enough sticks left to fill a mould.

Think together

1 Alex has baked 8 cupcakes.

She has 40 chocolate chips to share equally between the cupcakes.

How many chocolate chips can she use on each cupcake?

☐ ÷ 8 = ☐

She can use ☐ chocolate chips on each cupcake.

2 Use the diagrams to work out the divisions.

a) 72 ÷ 8 = ☐

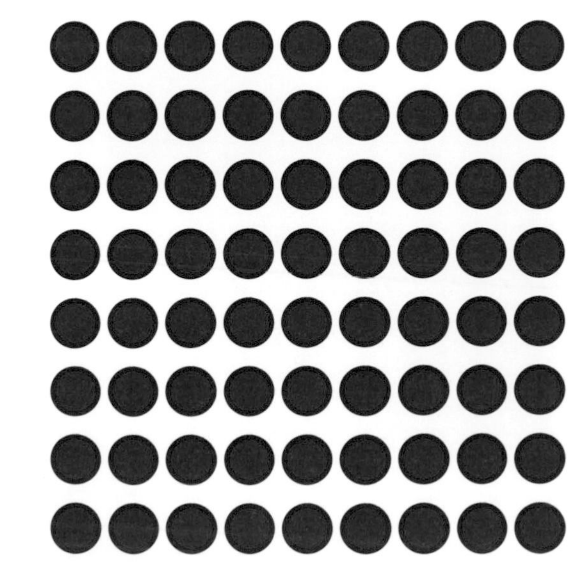

b) 48 ÷ 8 = ☐

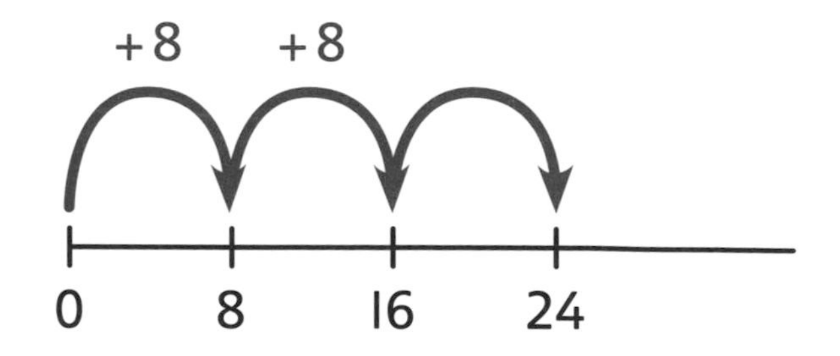

3 Lexi is cutting a cake.

First she cuts it in half.

She cuts each piece in half.

Then she cuts each new piece in half.

a) How can Lexi use what she has done to work out:

$16 \div 2$, $16 \div 4$ and $16 \div 8$?

b) Write a rule to divide by 8.

Use your rule to work out $88 \div 8$.

I used the candles to help me work out each calculation.

What is Lexi doing to each piece of cake each time? I think this will help me come up with a rule I can use.

→ Practice book 3A p127

8 times-table

Discover

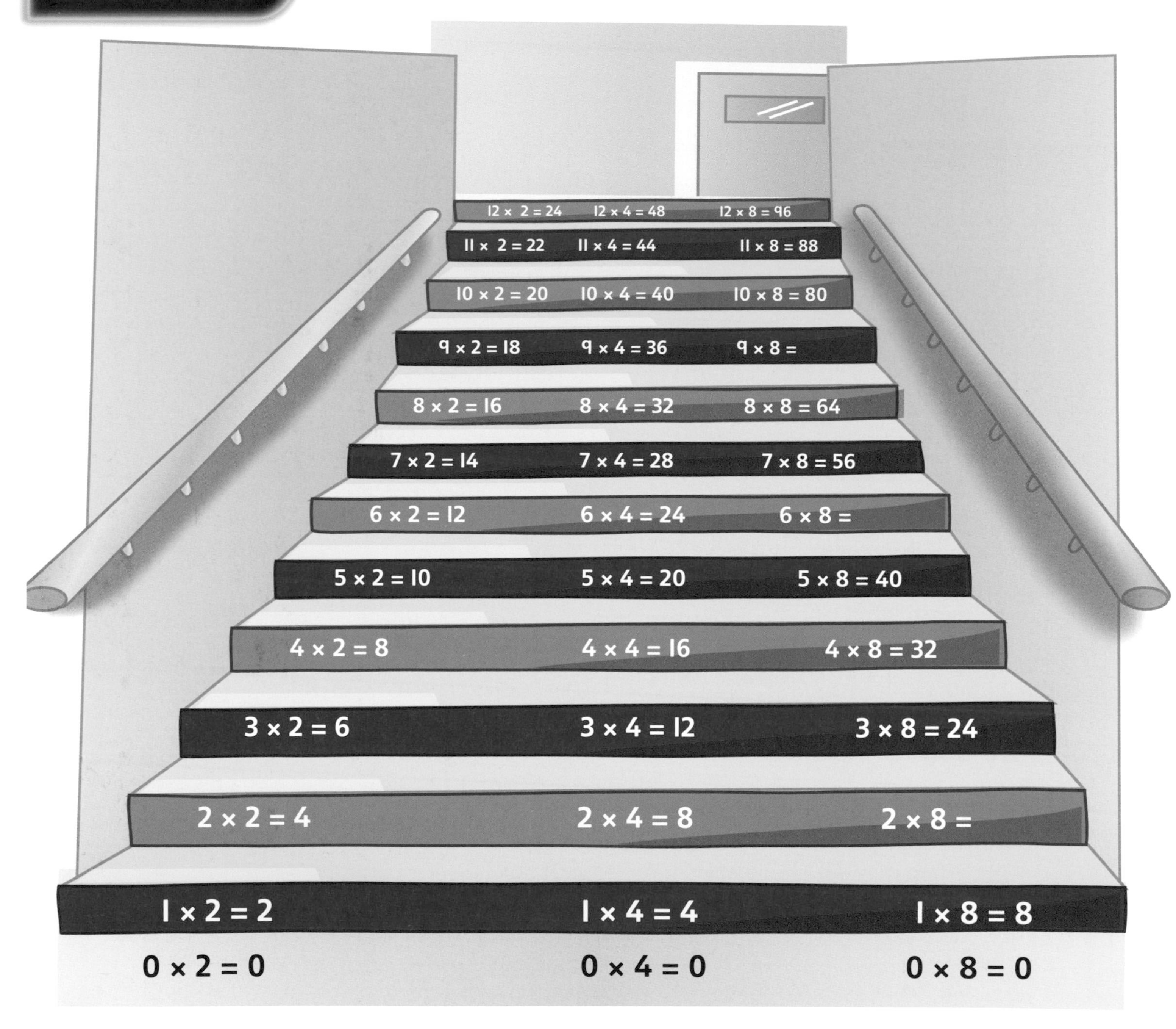

1 **a)** What answers are missing?

b) What is the connection between the 2, 4 and 8 times-table?

Share

a) You need to work out 2 × 8, 6 × 8 and 9 × 8.

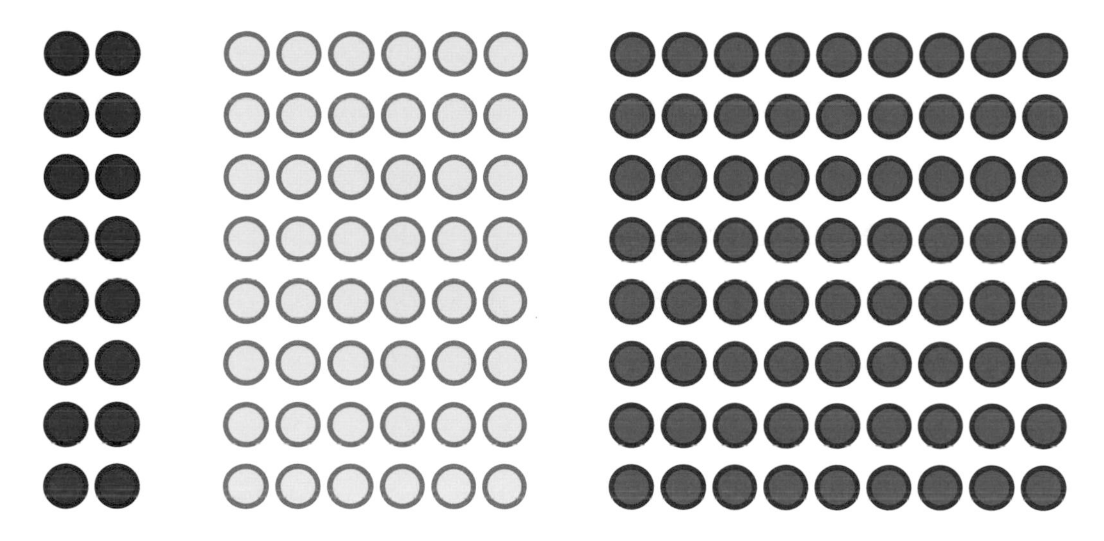

I used arrays to work out the answers. Now I need to try to remember these facts.

b) Compare the times-tables using arrays.

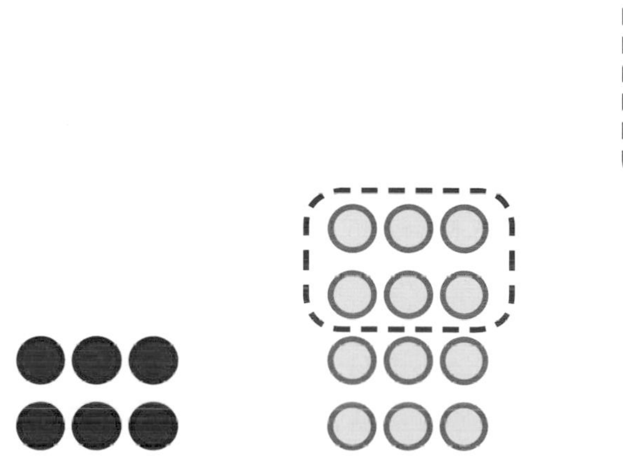

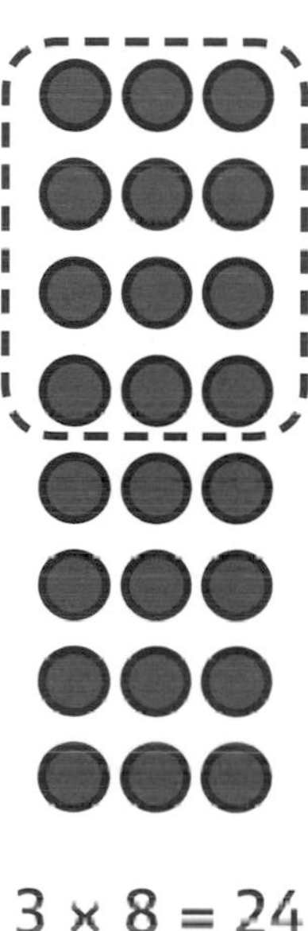

3 × 2 = 6   3 × 4 = 12   3 × 8 = 24

I can see a pattern. The answers double each time.

Think together

1 How many of each object are there in total?

What multiplication fact can you use to work out each total?

a)

☐ × ☐ = ☐

There are ☐ bottles of water.

b)

☐ × ☐ = ☐

There are ☐ cubes.

c)

☐ × ☐ = ☐

There are ☐ eggs.

2 How many of these do you know off by heart?

a) 3 × 8 = ☐

b) 10 × 8 = ☐

c) ☐ × 8 = 8

d) ☐ = 12 × 8

e) 40 ÷ 8 = ☐

f) 5 × 8 = ☐

g) 8 × 2 = ☐

h) ☐ ÷ 8 = 0

3 Find all the matching answers.

What is the pattern?

5 × 8	2 × 4	12 × 2
3 × 8	4 × 20	32 × 2
1 × 8	10 × 4	2 × 40
8 × 11	4 × 22	4 × 2
8 × 10	6 × 4	2 × 44
8 × 8	16 × 4	20 × 2

I have found some matches without working out any of the calculations.

I remember that 4 × 22 is the same as 2 × 44.

→ Practice book 3A p130

Problem solving – multiplication and division 1

Discover

1 **a)** The plants are planted in rows of 4.

There are 24 to plant.

How many rows of plants will there be?

b) Amal has some bunches of flowers.

There are 10 flowers in each bunch.

How many flowers does he have in total?

Share

a) There are 24 plants.

I will put them into groups of 4.

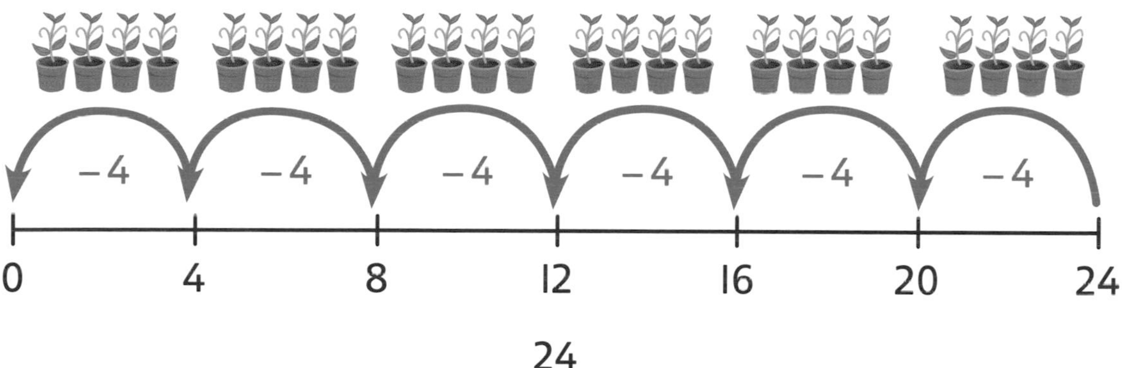

24

4	4	4	4	4	4

$24 \div 4 = 6$

There will be 6 rows of 4 plants.

I put them into groups of 4 using a bar model. I added bars until I got to 24.

b) Amal has 3 bunches of flowers.

$3 \times 10 = 30$

10	10	10

There are 10 flowers in each bunch.

$3 \times 10 = 30$

Amal has 30 flowers in total.

We can use bar models to help us see when to divide and when to multiply.

Think together

1. How many plants are there in total?

I think there are two multiplications you could do. I wonder what the bar model would look like for each one.

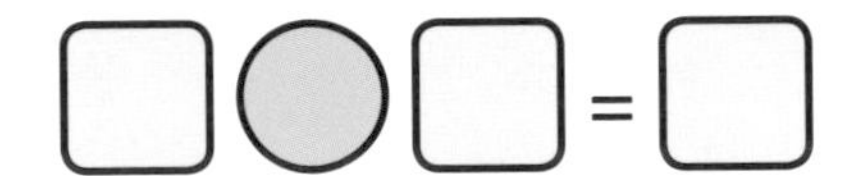

There are ☐ plants.

2. 30 flowers are shared equally between the 5 vases.

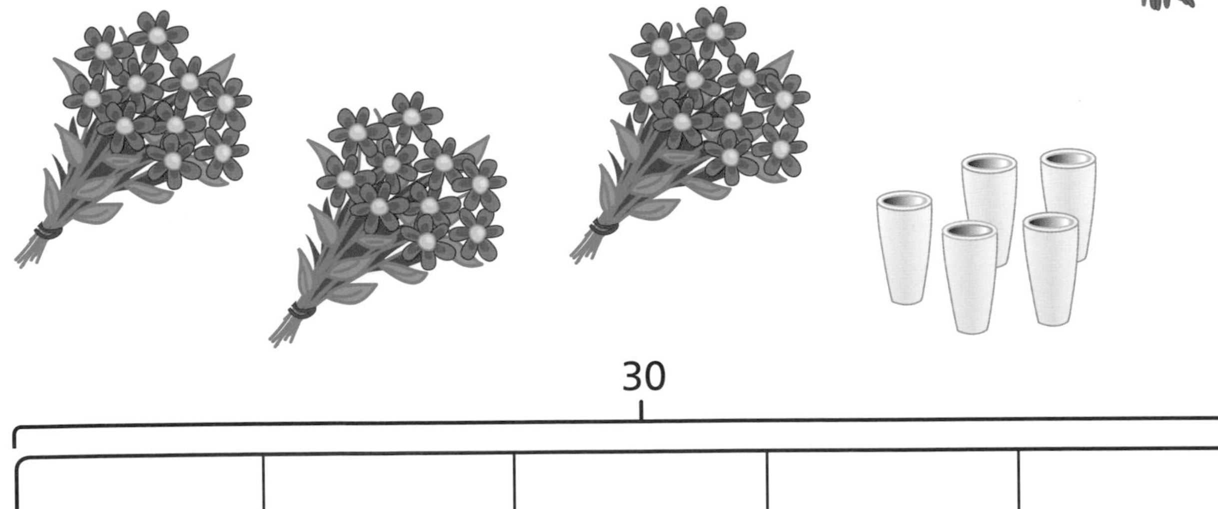

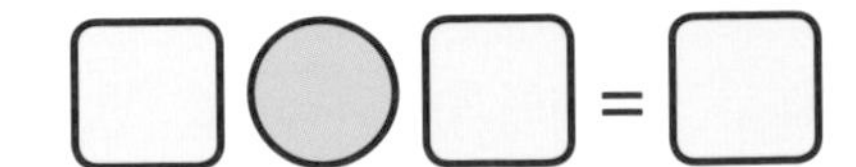

There are ☐ flowers in each vase.

3 Clare buys 8 small plant pots and 4 large plant pots.

The 4 large plant pots cost the same as 8 small plant pots.

How much does a large plant pot cost?

Write down all your steps.

I think this question involves a multiplication and a division. I will need to draw two bar models.

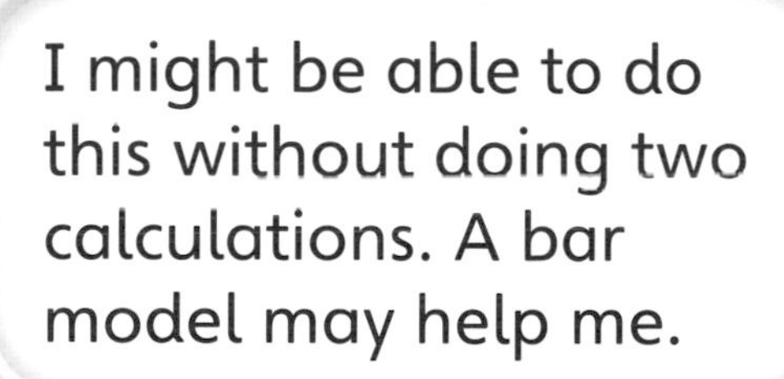

I might be able to do this without doing two calculations. A bar model may help me.

→ Practice book 3A p133

Problem solving – multiplication and division 2

Discover

1 **a)** Andy has put 3 blocks end to end to make a new shape.

What is the length of Andy's shape?

b) Isla makes a shape that is 32 cm long.

How many blocks does she use?

Share

a) Andy puts down 3 blocks.

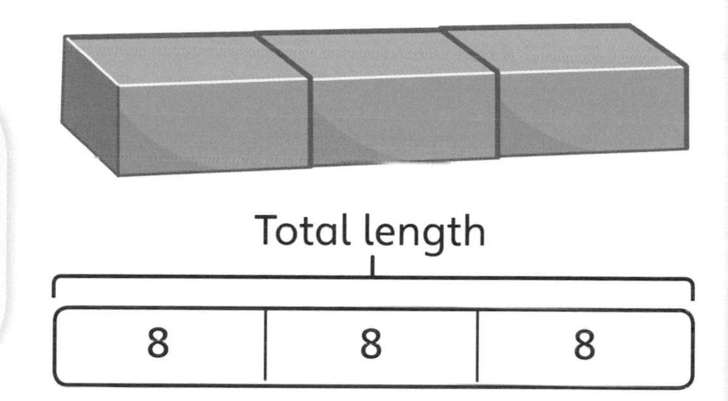

I will use a bar model to help me see what I should do.

I need to multiply to work out the total length.

$3 \times 8 = 24$

The length of Andy's shape is 24 cm.

b) Isla's shape is 32 cm long.

$4 \times 8 = 32$

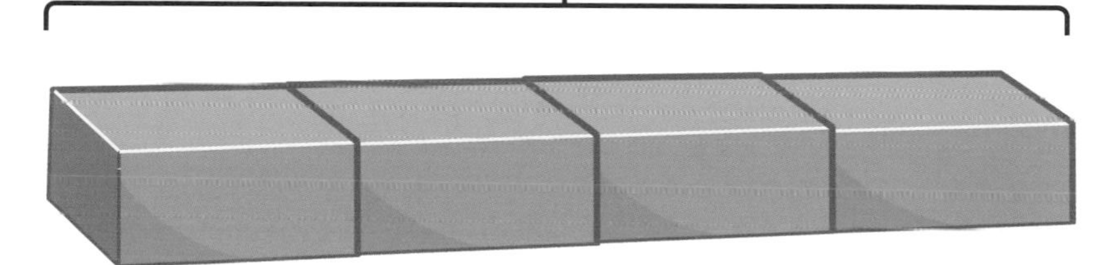

I laid the blocks down and kept adding on until I got to 32 cm.

I used division. I think this is a quicker way.

$32 \div 8 = 4$

Isla uses 4 blocks.

There is another possible answer:

$32 \div 4 = 8$

Isla uses 8 blocks.

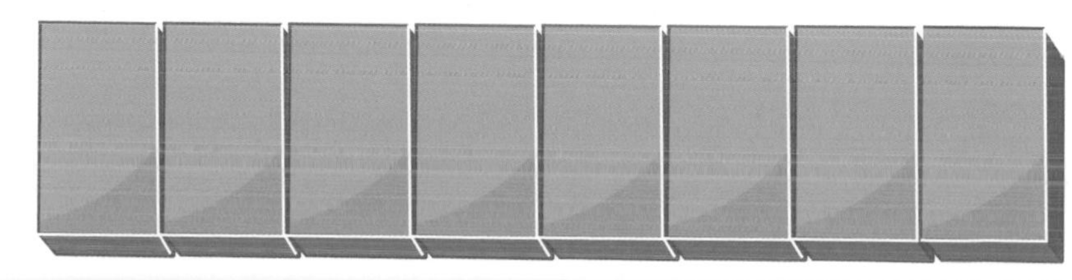

Think together

1 How long is the new shape that has been made?

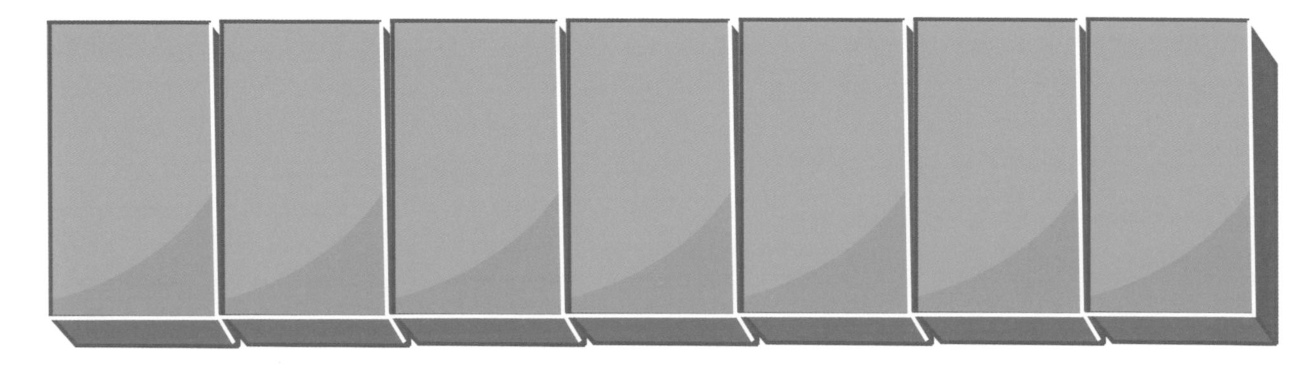

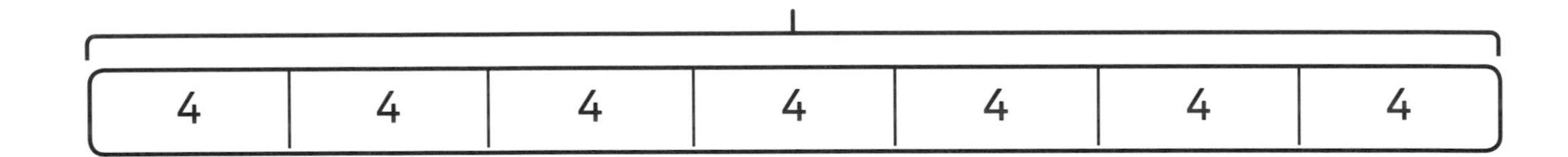

[ ] × [ ] = [ ]

The shape is [ ] cm long.

2 Which tower is taller?

How much taller is it?

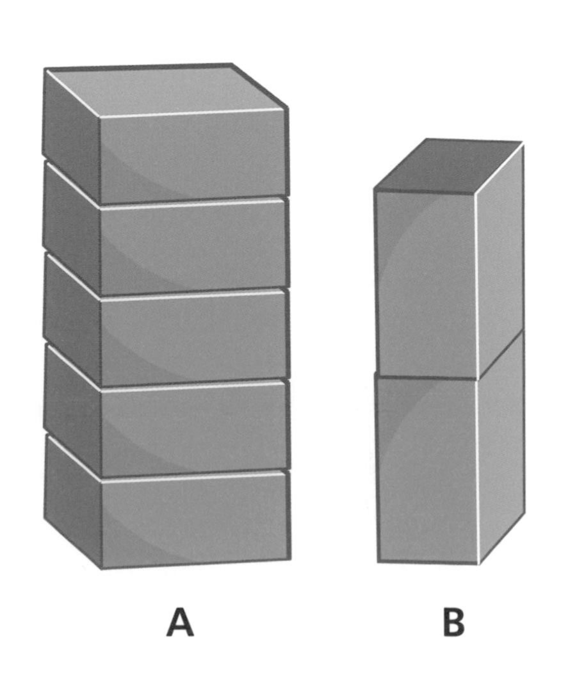

Tower A: [ ] × [ ] = [ ] cm tall.

Tower B: [ ] × [ ] = [ ] cm tall.

Tower [ ] is the tallest tower.

It is [ ] cm taller.

3 Isla makes this pattern using 7 wooden blocks.

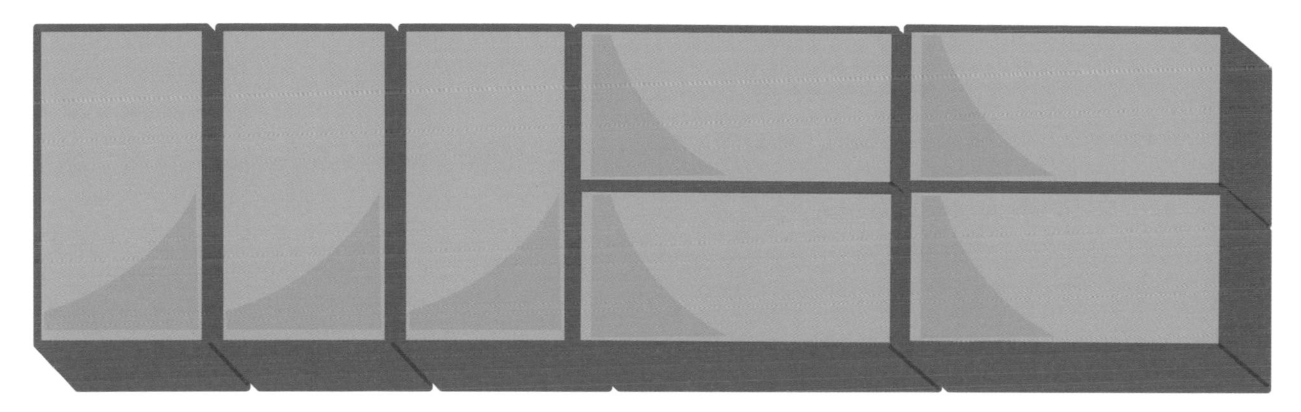

How long is the new pattern?

I think I need to work out two multiplications and then add.

These are the same blocks that the children used in the Discover activity.

→ Practice book 3A p136

Understanding divisibility 1

Discover

1 **a)** Lexi and Zac are using lollipop sticks to make squares.

How many squares can they make?

How many lollipop sticks are left over?

b) How would the answer change if they had 14 lollipop sticks?

What about 15, 16 or 17 lollipop sticks?

Share

We call the amount left over the **remainder**.

a) Four lollipop sticks make one square.

They can make 3 squares with 1 lollipop stick left over.

I will try organising my work in a table.

b)

Number of sticks	Working	Number of squares	Number of sticks left over
14		3	2
15		3	3
16		4	0
17		4	1

Think together

1 Lexi and Zac use more lollipop sticks.

How would you complete the table?

Number of sticks	Working	Number of squares	Number of sticks left over
18		4	
19			
20			

2 **a)** Describe the pattern that Lexi can see.

I can see a pattern in the number of lollipop sticks left over.

Lexi

b) Is Zac correct?

I don't think you can have more than 3 lollipop sticks left over.

Zac

3 Lexi and Zac are now making triangles using lollipop sticks.

Complete the table.

Number of sticks	Working	Number of triangles	Number of sticks left over
10		3	1
11			
12			
13			
14			
15			

There is a similar pattern to last time.

I wonder what the greatest number of lollipop sticks you can have left over is.

→ Practice book 3A p139

Understanding divisibility 2

Discover

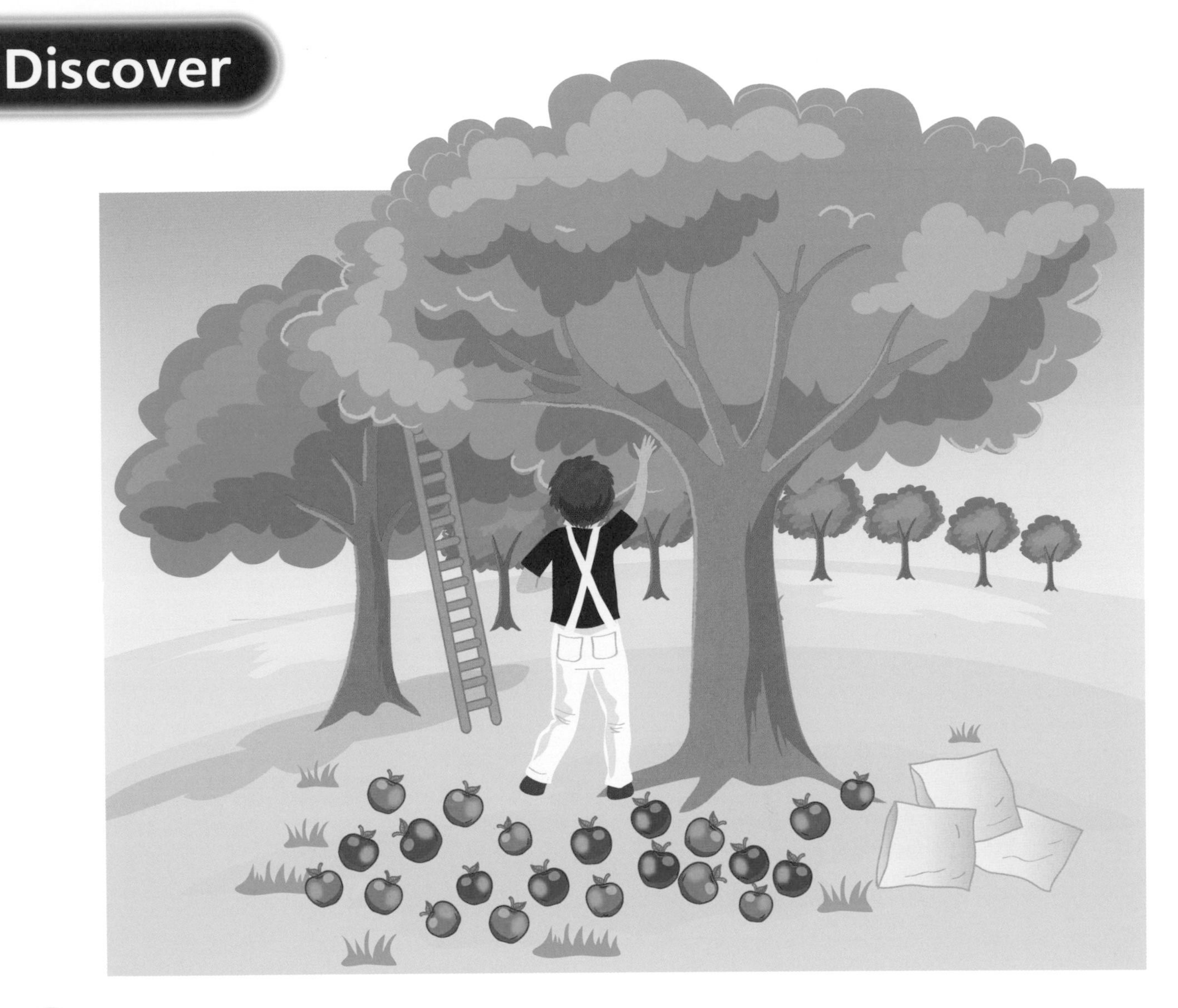

1 **a)** There are 22 apples

They are packed in bags of 5.

How many full bags are made?

How many apples are left over?

b) Write the calculation as a division.

☐ ÷ ☐ = ☐ remainder ☐

Share

a) Start with 22 apples.

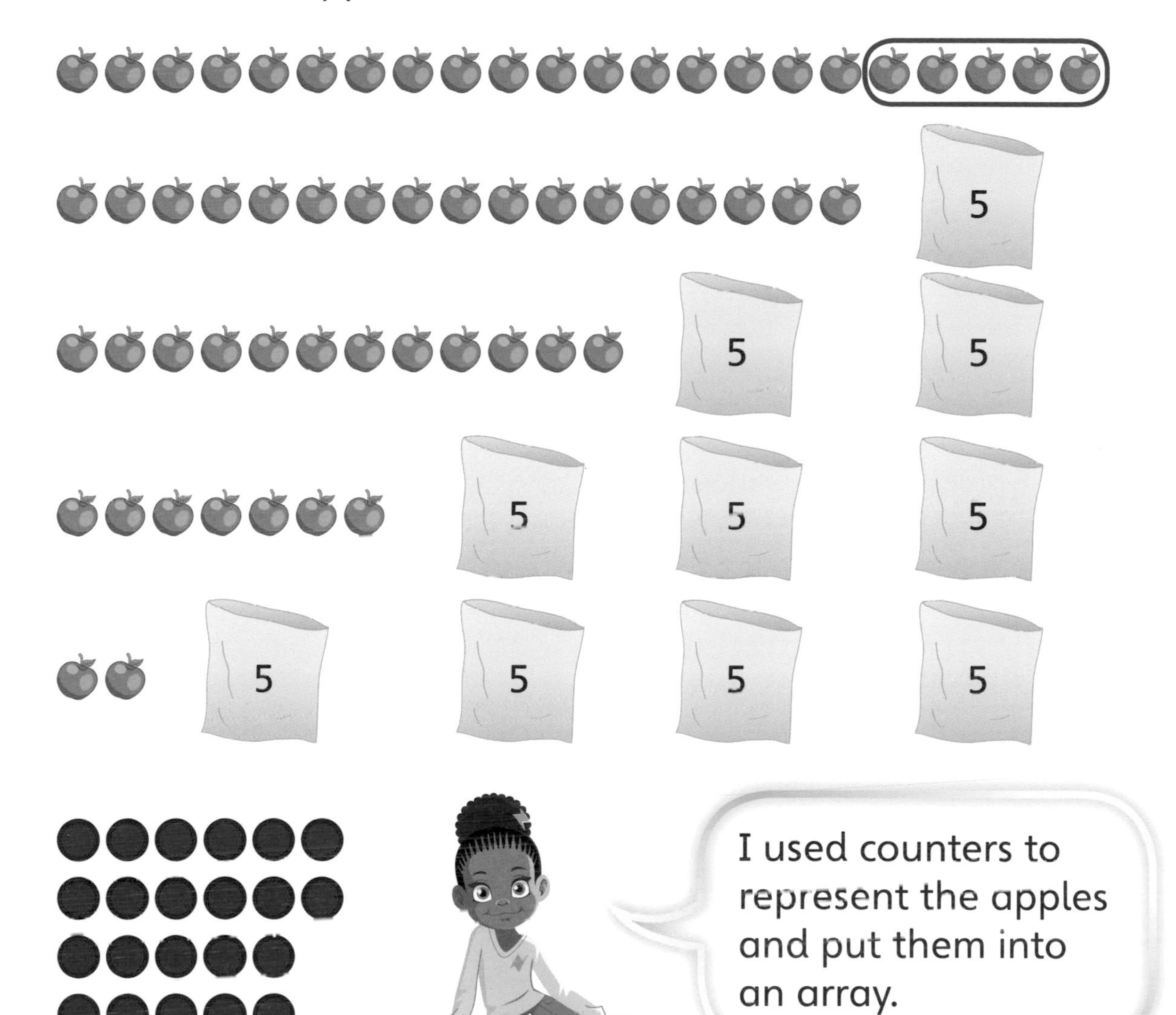

I used counters to represent the apples and put them into an array.

4 full bags are made and 2 apples are left over.

b) This is grouping.

There are 22 apples. There are 5 apples in each bag.

There are 4 full bags and 2 apples left over.

$22 \div 5 = 4$ remainder 2

Think together

1 Here are some oranges.

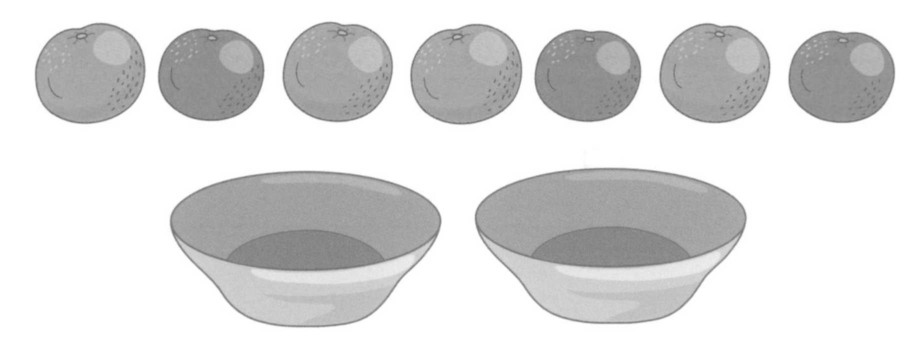

The oranges are shared between 2 bowls.

Can they be shared equally?

There are ☐ oranges in each bowl.

There are ☐ oranges left over.

☐ ÷ ☐ = ☐ remainder ☐

2 14 cubes are put into towers of 3.

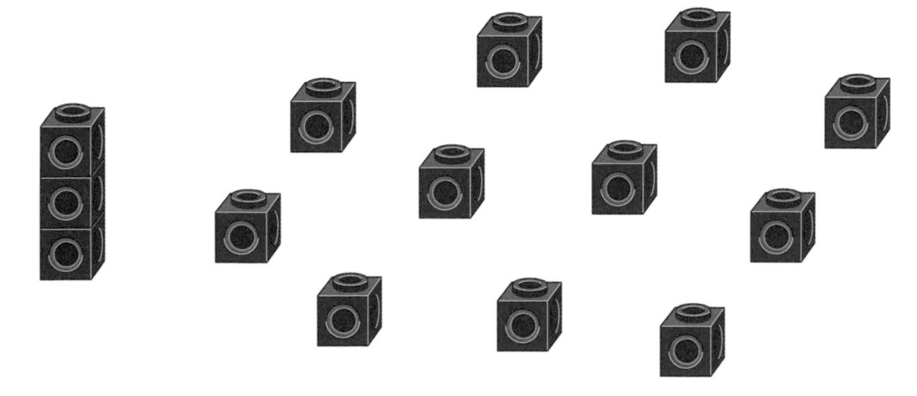

How many complete towers can be made?

How many cubes are left over?

☐ ÷ ☐ = ☐ remainder ☐

CHALLENGE

3 Explore the division calculations.

a) For each one, how many wholes are there and what is the remainder?

12 ÷ 5 = ☐ remainder ☐	17 ÷ 4 = ☐ remainder ☐
13 ÷ 8 = ☐ remainder ☐	51 ÷ 10 = ☐ remainder ☐

I know 2 × 5 = 10 and I know 3 × 5 = 15. This will help me work out 12 ÷ 5.

I wonder if I can use times-tables to help solve these.

b) Alex divides a number by 3. There is no remainder.

What could the number be? Where have you seen these numbers before?

→ Practice book 3A p142

Related facts – multiplication and division

Discover

1 **a)** What multiplication and **division facts** can you see?

☐ × ☐ = ☐ ☐ ÷ ☐ = ☐

☐ × ☐ = ☐ ☐ ÷ ☐ = ☐

b) Use counters to show $8 \times 3 = 24$

What else does this show?

Share

a)

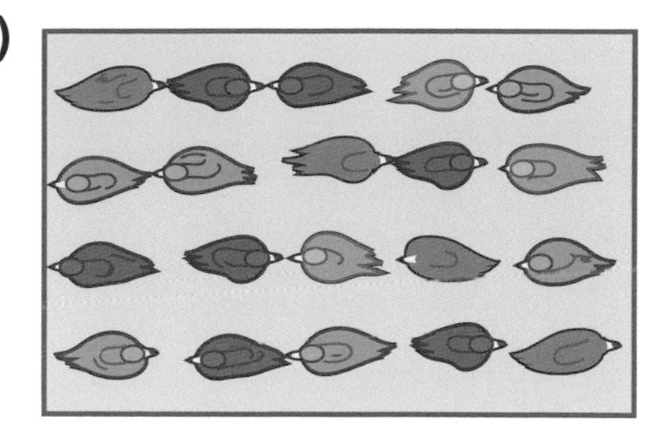

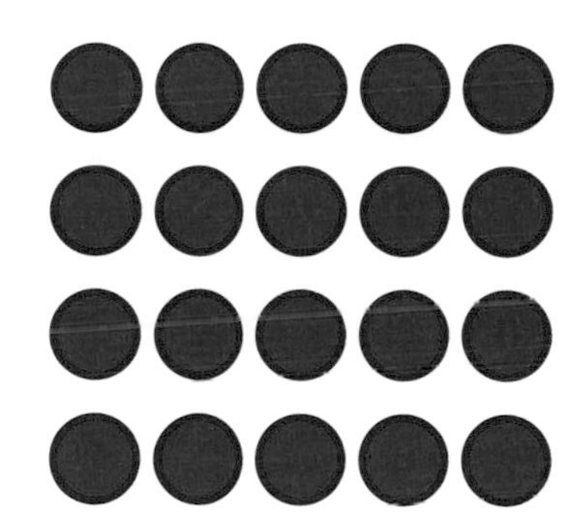

It looks like the flamingos are standing in an array.

There are 4 rows.

There are 5 columns.

There are 20 flamingos in total.

$4 \times 5 = 20$

An array shows four facts: two multiplication and two division facts.

$5 \times 4 = 20$

$20 \div 4 = 5$

$20 \div 5 = 4$

I think there are four more facts. What if you wrote the answer first?

b) This array shows

$8 \times 3 = 24$

$3 \times 8 = 24$

$24 \div 3 = 8$

$24 \div 8 = 3$

Think together

1 30 trees are planted in a forest.

They are planted in 3 rows.

There are 10 trees in each row.

Write down four facts that are shown by the trees.

$\square \times \square = \square$ $\qquad$ $\square \div \square = \square$

$\square \times \square = \square$ $\qquad$ $\square \div \square = \square$

2 32 cakes are arranged in 4 rows.

There are 8 cakes in each row.

Match each calculation to the correct statement.

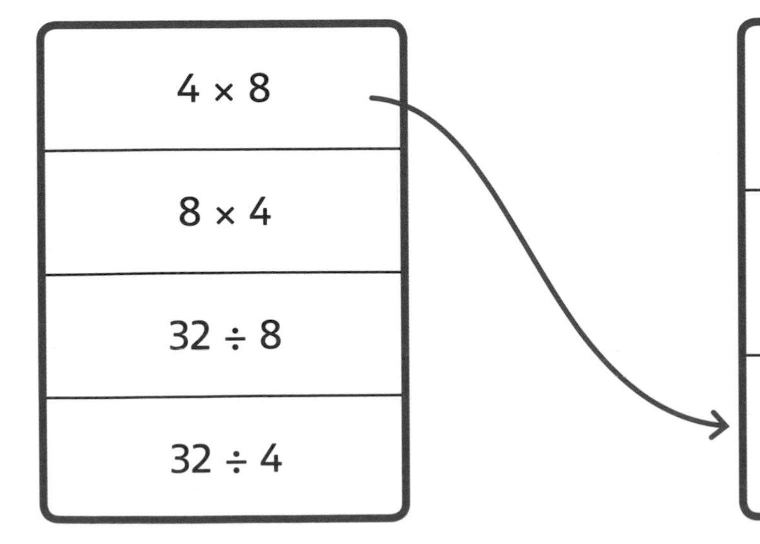

This calculation works out the number of rows.
This calculation works out how many cakes are in a row.
This calculation works out the total number of cakes.

3 Here is a trio – a group of 3 numbers.

a) Explain the connection between the numbers.

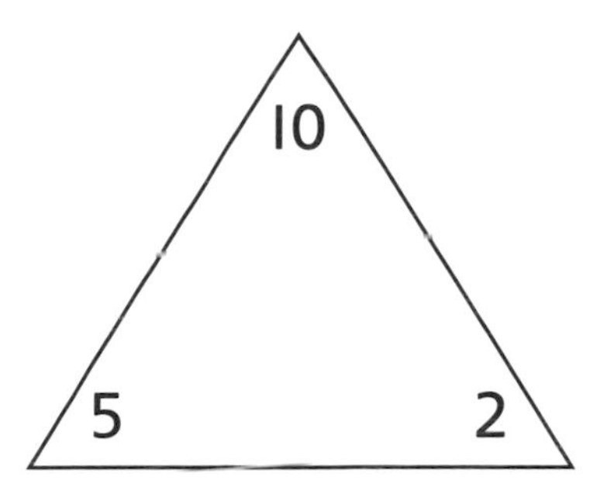

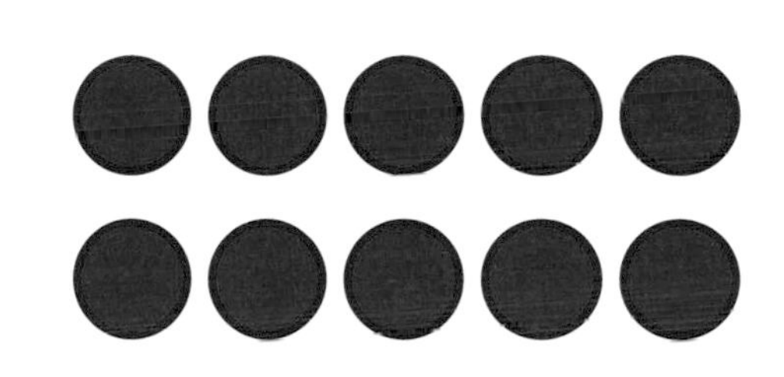

b) Can you work out the missing numbers in these trios?

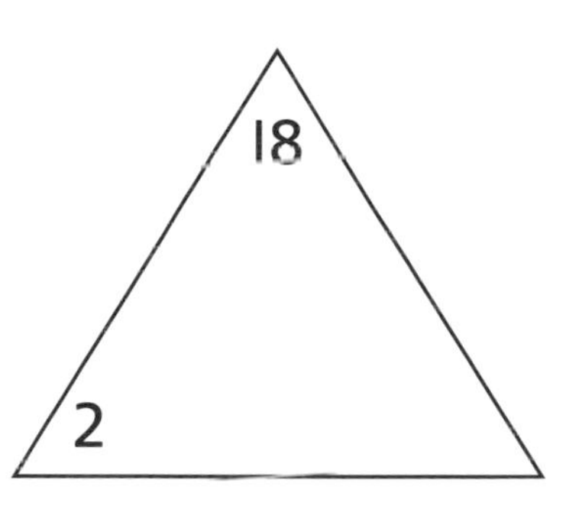

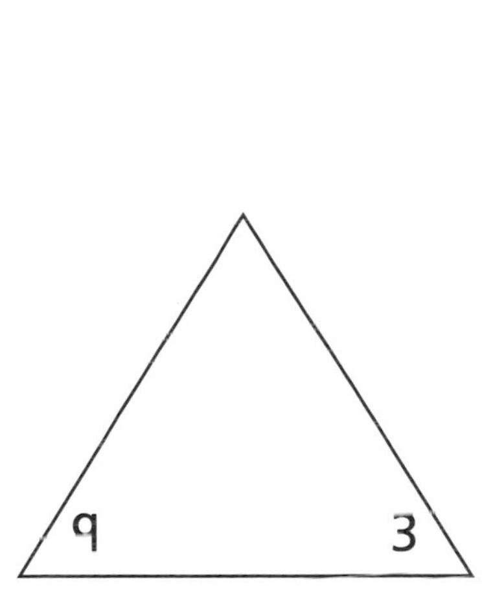

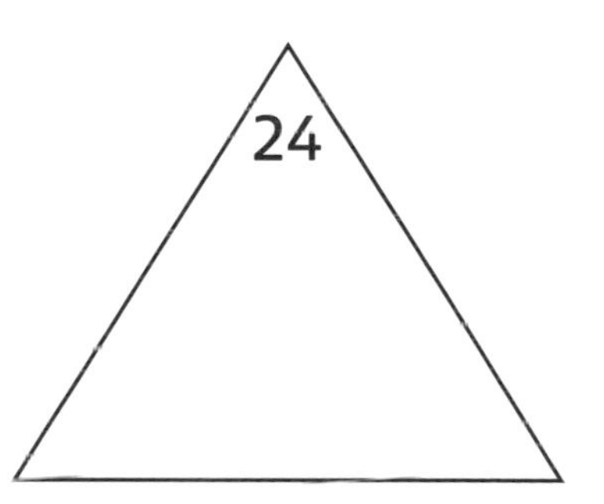

I will use counters to help me make an array.

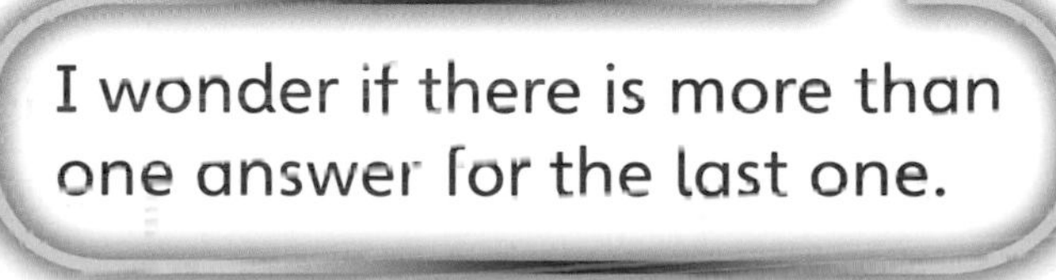

I wonder if there is more than one answer for the last one.

→ Practice book 3A p145

End of unit check

1 David shares 24 grapes between 3 people.

How many grapes does each person get?

A 21 B 6 C 8 D 72

2 Which calculation will not work out the number of counters in the array?

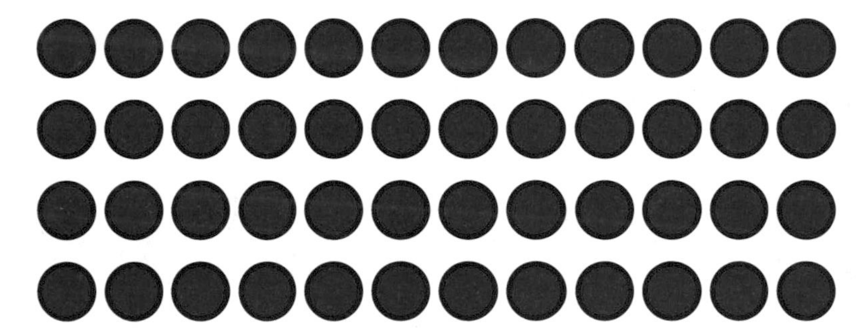

A 12 + 12 + 12 + 12

B 12 × 4

C 4 × 12

D 4 + 4 + 4 + 4

3 Which calculation gives the greatest answer?

A 7 × 3 B 8 × 2 C 6 × 4 D 0 × 10

4 Which calculation gives the same answer as 6×8?

A 3×8

B 12×4

C 9×5

D 12×16

5 A pack contains 4 bread rolls.

How many bread rolls are there in 7 packs?

A 11

B 35

C 28

D 24

6 Lexi shares 16 cubes equally between 3 people.

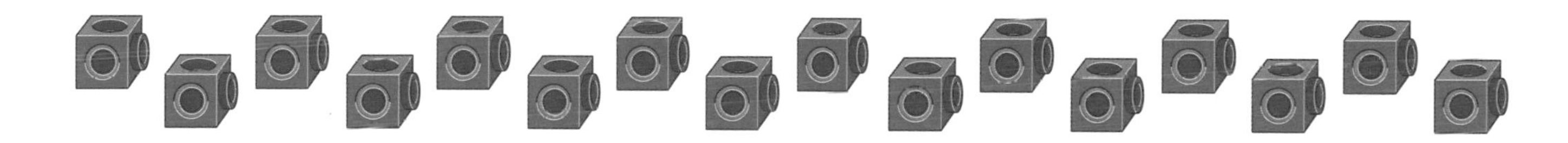

How many cubes do they each get? How many cubes are left over?

A 1 cube, 5 left over

B 4 cubes, 3 left over

C 6 cubes, 0 left over

D 5 cubes, 1 left over

7 What is the missing value?

$\square \times 4 = 24 \div 3$

→ Practice book 3A p148

Practice helps us get better!

I enjoyed finding new methods!

Wow, we have solved
some difficult problems!

Yes, we have! Can we find even better ways to solve problems?

We have learnt lots to help us next term.

I am looking forward to the challenge of the next book.

What do we know now?

Can you do all these things?

- Count in 100s
- Use the number line up to 1,000
- Find 1, 10 or 100 more or less
- Compare numbers to 1,000
- Count in 50s

Some of it was difficult, but we did not give up!

Now you are ready for the next books!

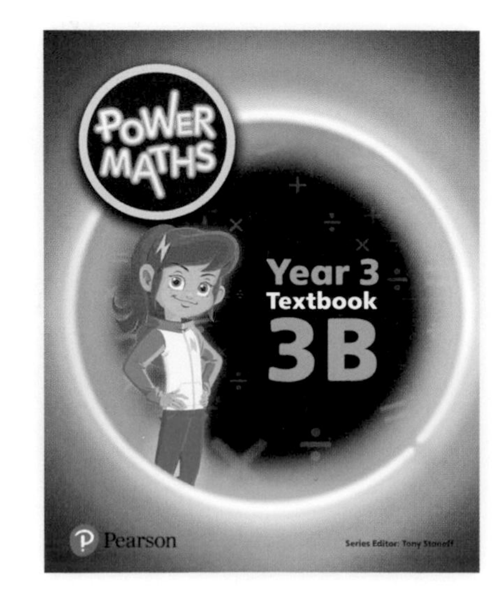